Power IC Design

With insight & intuition

1st Edition

By

Gabriel Alfonso Rincón-Mora

School of Electrical and Computer Engineering
Georgia Institute of Technology

Rincon-Mora.gatech.edu

V. 1

Printed in the United States of America.

Discovering the universe through the art of design

Contents

Preface

I. Intended Audience

Integrated-circuit (IC) design, system, and product engineers, managers, and undergraduate and graduate engineering students engaged or interested in expanding their knowledge on how to analyze, design, evaluate, specify, develop, and test power supply ICs.

II. Description and Objectives

This slide book uses design insight, real-life examples, illustrative figures, easy-to-follow equations, and simple SPICE code to show how semiconductor devices (diodes, bipolar-junction transistors (BJTs), and metal–oxide–semiconductor (MOS) field-effect transistors (FETs)) work independently and collectively in switched-inductor power supplies; how these power supplies transfer power, consume power, and react and respond across frequency; how feedback loops switch, control, and stabilize them; and how the building blocks that comprise them are implemented and designed.

G.A.R.M.

Atlanta, Georgia

Chapter 1. Power-Supply Systems

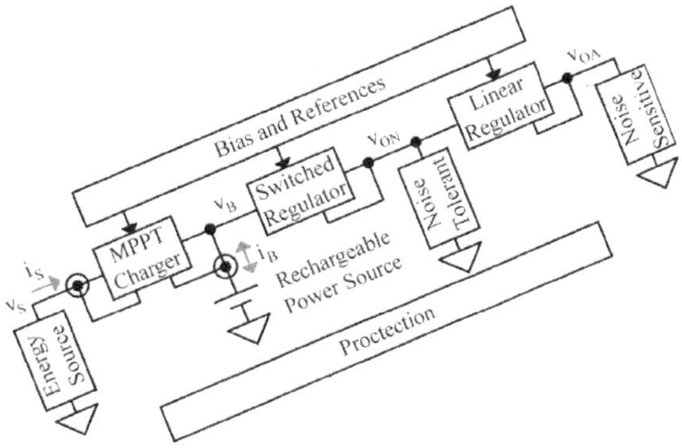

1.1. Modern Applications

Bio-Monitors

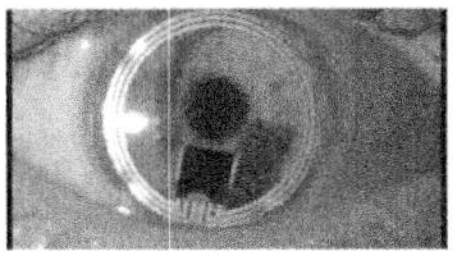

μ-Sensors

10–200 μW

Pacemakers

Cochlear Implants

Hearing Aids

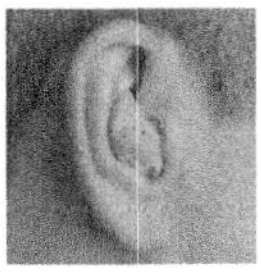

Surveillance

Tele-meters

μ-Robots

Neural Recorders

0.2–10 mW

Neural Stimulators

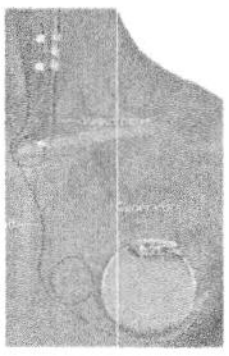

Retinal Implants

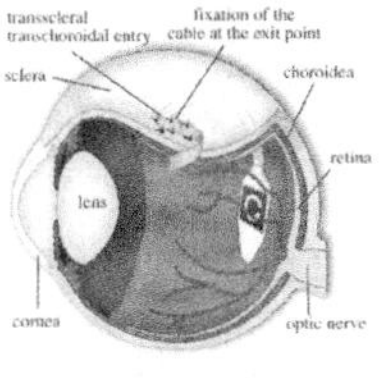

0.1–3 W

Smart Phones

…

Common Requirements

Useful: Sense, process, transmit

Unobtrusive: Small, wireless

Economical: Silicon microchips

1.2. Power Transfer

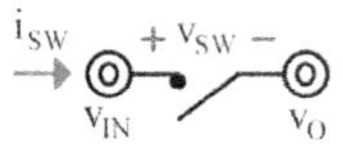

Switch: $i_{SW} \propto P_{IN/O}$

$P_R = i_{SW} v_{SW}$

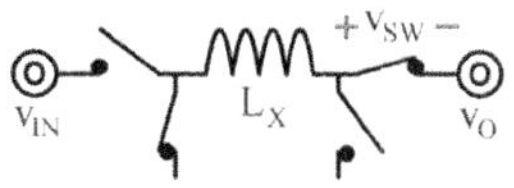

Linear: $v_{SW} = v_{IN} - v_O = 0.2\text{–}5\ V \neq f(i_{SW})$

Switched: $v_{SW} = i_{SW} R_{SW} \approx 10\text{–}50\ mV$

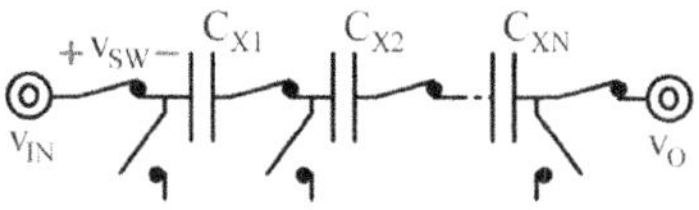

∴ Switched P_R < Linear P_R

Switched Network: Switched inductor → 2 to 4 switches per transfer

Switched capacitors → Normally, more switches per transfer

∴ SC's usually require more gate-charge power P_G

Switched Inductor: $P_R = i_L R_L$ → Low R_L } Bulky L_X

→ Low i_L → High L_X

∴ Use only one off-chip, in-package, or on-chip L_X

1.3. System Composition

Elect. Systems: Require steady supplies → Voltage regulators

Volt. Regulators: Linear → Supply P_O, regulate v_O, suppress noise v_{no}, lose P_R

Switched → Supply P_O, regulate v_O, generate v_{no}, lose little P_R

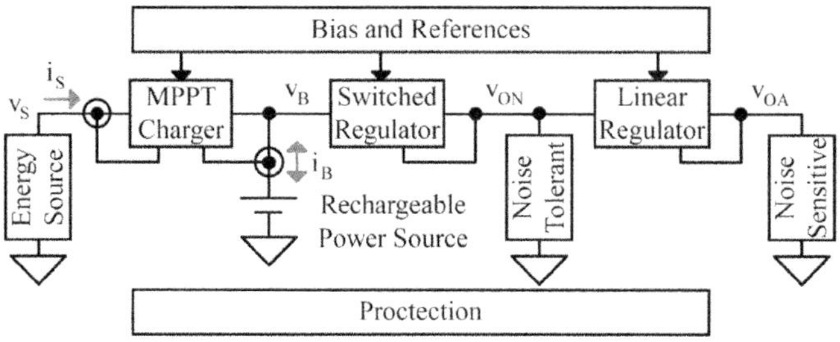

Noise-Sensitive Loads: Require low-noise supplies → Use linear regulator

Noise-Tolerant Loads: Survive switching noise → Use switched regulator

Battery: Exhaustible-E_B, variable-v_B power source → Use switched charger

1.4. Electrical-Engineering Foundation

Laplace: $s = i\,(2\pi f_O) \propto f_O$ $\qquad i^2 = -1$

Impedances: $R_X \neq f(f_O)$ $\qquad sL_X \propto f_O$ $\qquad \dfrac{1}{sC_X} \propto \dfrac{1}{f_O}$

Ohm's Law: $v_X = i_X Z_X$ $\qquad i_X = \dfrac{v_X}{Z_X}$ $\qquad P_R = i_R v_R = i_R^{\,2} R_X = \dfrac{v_R^{\,2}}{R_X}$

Parallel Z's:

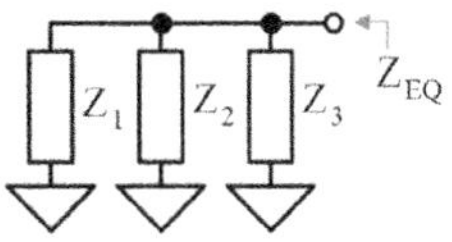

$$Z_{EQ} = Z_1 \parallel Z_2 \parallel Z_3 = \frac{Z_1 Z_2 Z_3}{Z_1 Z_2 + Z_1 Z_3 + Z_2 Z_3}$$

$$\leq \mathrm{Min}\{Z_1, Z_2, Z_3\}$$

Voltage Divider:

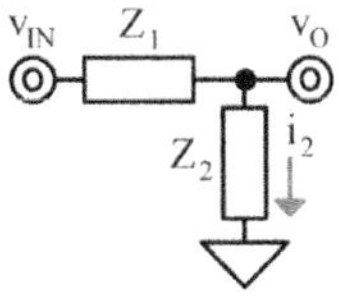

$$v_O = i_2 Z_2 = \left(\frac{v_{IN}}{Z_1 + Z_2}\right) Z_2 \qquad A_V \equiv \frac{v_O}{v_{IN}} = \frac{Z_2}{Z_1 + Z_2}$$

Diode: One-way conductor

Transistor: Controllable resistor

Chapter 2. Power Devices

2.1. Diodes: Large Signal

Response Time

2.2. MOSFETs: N Channel

P Channel

Capacitances

MOS Diodes

Junction Isolation

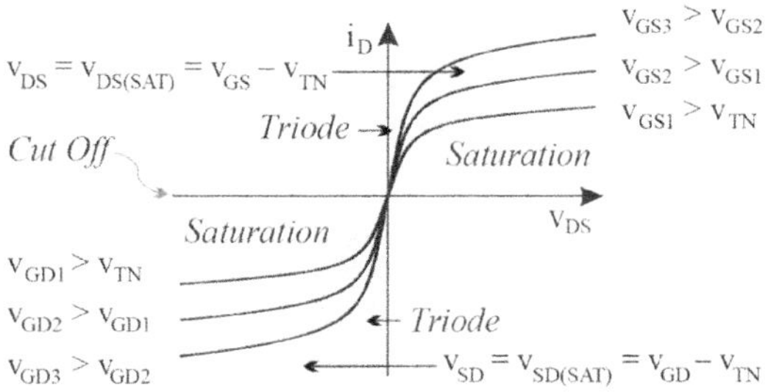

2.1. Diodes: A. Large Signal

Regions

Forward Bias: $v_D > 0$

Zero Bias: $v_D = 0$

Reverse: $v_D < 0$

Breakdown: $v_D \approx -V_{BD}$

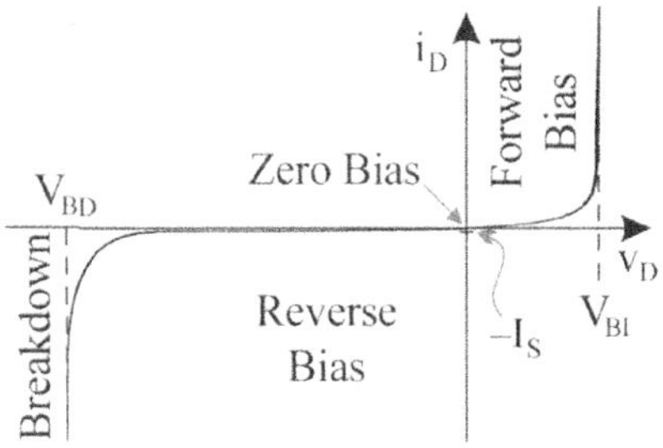

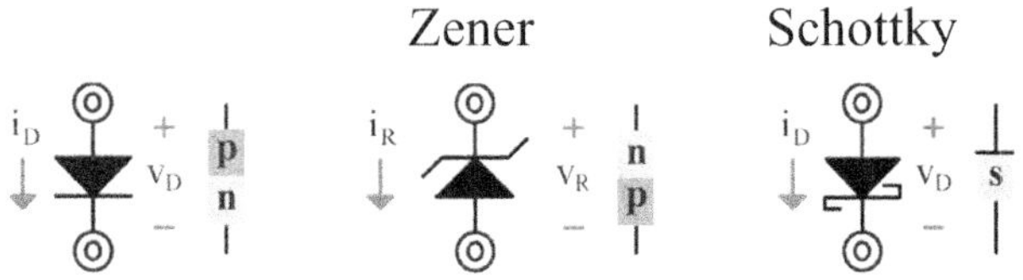

Parameters:

Built-in Voltage $\equiv V_{BI} \approx 600$ mV

Reverse-Saturation Current $\equiv I_S$

Thermal Voltage $\equiv V_t \approx 26$ mV at 27°C

Junction Area $\equiv A_J$

$$i_D = I_S\left[\exp\left(\frac{v_D}{V_t}\right) - 1\right]$$

$$\approx I_S \exp\left(\frac{v_D}{V_t}\right) \propto A_J$$

↑

When $v_D > 3V_t$

B. Response Time

Zero-v_D Junction Capacitance $\equiv C_{J0}$

C_{J0} per unit Area $\equiv C_{J0}''$

Transit Time $\equiv \tau_F$

Small Variations:

$$C_J = \frac{\Delta q_D}{\Delta v_D} = \frac{\Delta i_D \Delta t_R}{\Delta v_D} = C_{DEP} + C_{DIF}$$

$$C_{DEP} = \frac{C_{J0}}{\sqrt{\frac{V_{BI} - v_D}{V_{BI}}}} = \frac{A_J C_{J0}''}{\sqrt{1 + \frac{v_R}{V_{BI}}}} \propto \frac{A_J}{d_W}$$

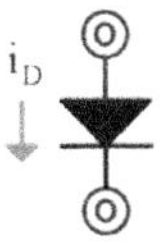

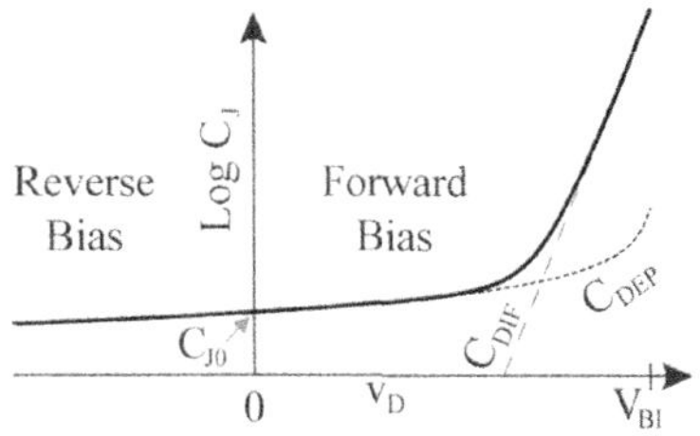

$$C_{DIF} = \frac{\Delta q_{DIF}}{\Delta v_D} = \frac{\Delta i_D \tau_T}{\Delta v_D} \approx \left(\frac{\partial i_D}{\partial v_D}\right)\tau_T \approx \left(\frac{I_D}{V_t}\right)\tau_T \equiv g_m \tau_T$$

Large Variations: Forward/Reverse Recovery

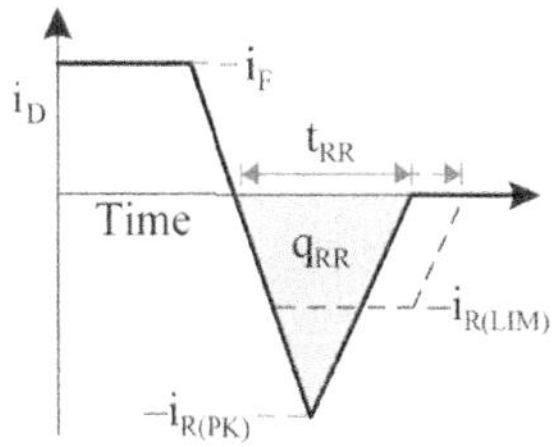

Feed $q_{DIF} + q_{DEP} \approx q_{DIF}$ $\therefore$ $t_{FR} = \frac{q_{FR}}{i_F} \approx \frac{q_{DIF}}{i_F} \approx \frac{i_F \tau_T}{i_F} = \tau_T$

Pull $q_{DIF} + q_{DEP} \approx q_{DIF}$ $\therefore$ $t_{RR} = \frac{q_{RR}}{i_R} \approx \frac{q_{DIF}}{i_R} \approx \left(\frac{i_F}{i_R}\right)\tau_T$ →

Schottky: $V_{BI} \approx 300\text{–}800$ mV No q_{DIF} → Short t_{RR}

2.2. MOSFETs: A. N Channel

Regions

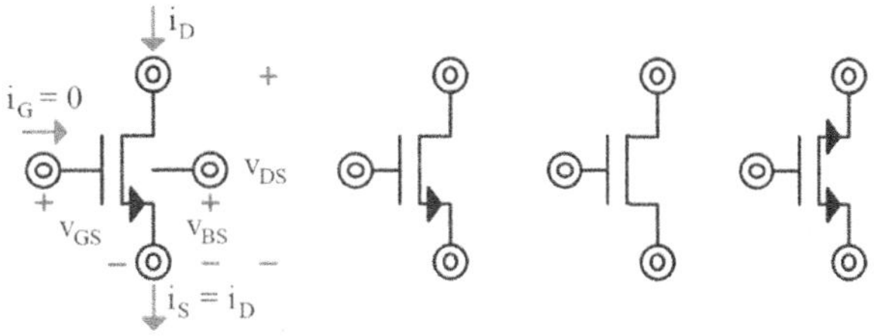

Accumulation $\equiv$ Cut Off: $v_{GS} < 0$

Depletion $\equiv$ Sub-v_T: $0 < v_{GS} < v_{TN}$

Triode: $v_{DS} < 3V_t$

$$i_D\Big|^{0<v_{GS}<v_{TN}} = \left(\frac{W}{L}\right) I_{SN} \exp\left(\frac{v_{GS} - v_{TN}}{n_I V_t}\right)\left[1 - \frac{1}{\exp(v_{DS}/V_t)}\right]$$

Saturation: $v_{DS} > v_{DS(SAT)}' \approx 3V_t$

$$i_D\Big|^{0<v_{GS}<v_{TN}}_{v_{DS}>3V_t} \approx \left(\frac{W}{L}\right) I_{SN} \exp\left(\frac{v_{GS} - v_{TN}}{n_I V_t}\right)$$

Parameters:

Channel Width & Length $\equiv$ W & L

Saturation Current $\equiv I_{SN}$

Threshold Voltage $\equiv v_{TN} \approx 400\text{–}600$ mV

Non-Ideality Factor $\equiv n_I \approx 1\text{–}2$

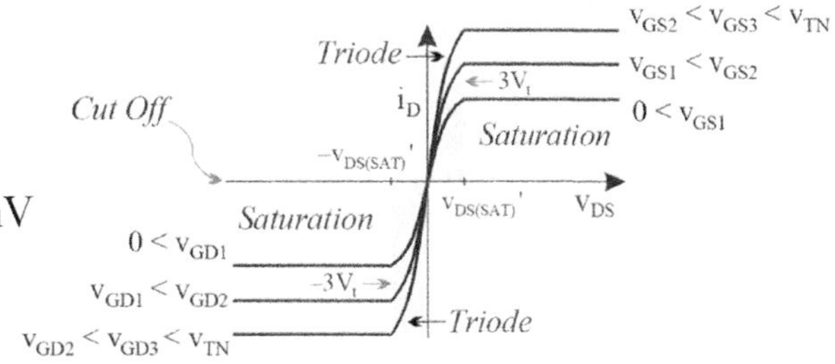

Inversion: $v_{GS} > v_{TN}$

Triode → $v_{DS} < v_{DS(SAT)}$

Saturation → $v_{DS} > v_{DS(SAT)}$

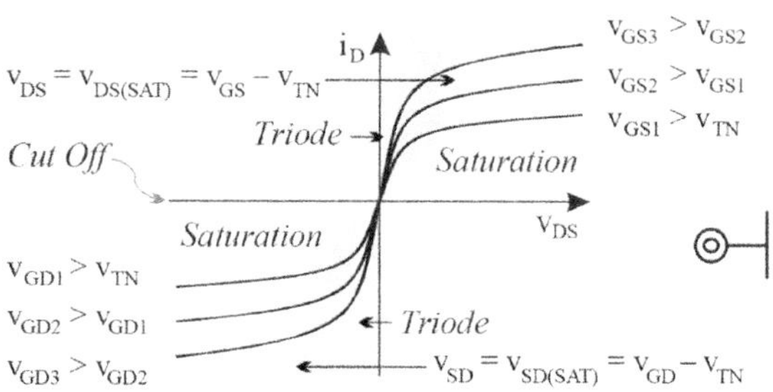

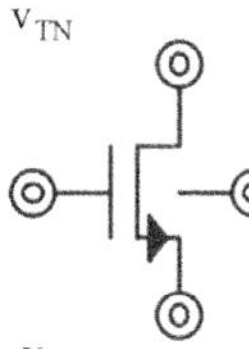

$$R_{CH}\Big|_{v_{DS} << v_{GST}}^{v_{GS} > v_{TN}} \approx \left(\frac{L}{W}\right)\left[\frac{1}{K_N'(v_{GS} - v_{TN})}\right]$$

$$v_{TN} = V_{TN0} + \gamma_N\left(\sqrt{2\psi_B - v_{BS}} - \sqrt{2\psi_B}\right)$$

Surface Barrier Potential $\equiv \psi_B$

$$i_D\Big|_{v_{DS} < v_{GST}}^{v_{GS} > v_{TN}} = \frac{v_{DS}}{R_{CH}} = v_{DS}\left(\frac{W}{L}\right)K_N'\left(v_{GS} - v_{TN} - \frac{v_{DS}}{2}\right)$$

$$K_N' = \mu_N C_{OX}'' = \mu_N\left(\frac{\varepsilon_{OX}}{t_{OX}}\right) = \mu_N\left(\frac{3.9\varepsilon_0}{t_{OX}}\right)$$

$$i_D\Big|_{v_{DS} > v_{GST}}^{v_{GS} > v_{TN}} = \frac{v_{DS(SAT)}}{R_{CH}} \approx \left(\frac{W}{L}\right)\left(\frac{K_N'}{2}\right)(v_{GS} - v_{TN})^2(1 + \lambda_N v_{DS})$$

$$v_{DS(SAT)} = v_{GS} - v_{TN} \equiv v_{GST}$$

Parameters: Zero-v_{BS} $v_{TN} \equiv V_{TN0}$ — Body-Effect Parameter $\equiv \gamma_N \approx 0.4$–$0.6\ \sqrt{V}$

Transconductance Parameter $\equiv K_N'$ — Channel-Length Modulation Parameter $\equiv \lambda_N$

Electron Mobility $\equiv \mu_N$ — Oxide Capacitance/Area $\equiv C_{OX}''$

Example: Determine R_{CH} when $W_{CH} = 10\ \mu m$, $L_{CH} = 120$ nm, $v_{GS} = 1.8$ V, $v_{DS} = 50$ mV, $\mu_N = 72k\ mm^2/V{\cdot}s$, $t_{OX} = 12.5$ nm, $v_{TN} = 400$ mV.

Solution: $$C_{OX}'' = \frac{\varepsilon_{OX}}{t_{OX}} = \frac{\varepsilon_{Si}\varepsilon_0}{t_{OX}} = \frac{3.9(8.845\ pF/m)}{12.5\ nm} = 2.76\ mF/m^2 = 2.76\ fF/\mu m^2$$

$$K_N' = \mu_N C_{OX}'' = (72m)(2.76m) = 200\ \mu A/V^2$$

$$v_{DS} = 50\ mV << v_{GST} = v_{GS} - v_{TN} = 1.8 - 400m = 1.4\ V$$

$$\therefore\ \text{Deep in Triode} \rightarrow R_{CH} \approx \left(\frac{L_{CH}}{W_{CH}}\right)\left[\frac{1}{K_N'(v_{GS} - v_{TN})}\right] = 43\ \Omega$$

Example: Determine v_{TN} when $V_{TN0} = 400$ mV, $v_{BS} = -100$ mV, $\gamma_N = 600\ m\sqrt{V}$, $\psi_B = 300$ mV.

Solution: $$v_{TN} = V_{TN0} + \gamma_N\left(\sqrt{2\psi_B - v_{BS}} - \sqrt{2\psi_B}\right)$$

$$= 400m + (600m)\left(\sqrt{2(300m) - (-100m)} - \sqrt{2(300m)}\right) = 440\ mV$$

B. P Channel

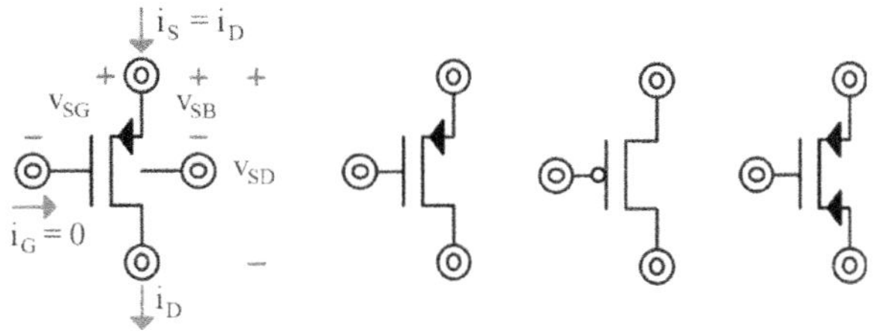

Regions $\quad v_G > v_S$ ↓

Accumulation ≡ Cut Off: $v_{SG} < 0$

Depletion ≡ Sub-v_T: Triode Saturation

Inversion: Triode Saturation

$$i_D\Big|^{0<v_{SG}<|v_{TP}|} = \left(\frac{W}{L}\right) I_{SP} \exp\left(\frac{v_{SG} - |v_{TP}|}{n_I V_t}\right)\left[1 - \frac{1}{\exp(v_{SD}/V_t)}\right]$$

$$R_{CH}\Big|^{v_{SG}>|v_{TP}|}_{v_{SD}<<v_{SGT}} \approx \left(\frac{L}{W}\right)\left[\frac{1}{\mu_P C_{OX}''(v_{SG} - |v_{TP}|)}\right]$$

$$i_D\Big|^{0<v_{SG}<|v_{TP}|}_{v_{SD}>3V_t} \approx \left(\frac{W}{L}\right) I_{SP} \exp\left(\frac{v_{SG} - |v_{TP}|}{n_I V_t}\right)$$

$$i_D\Big|^{v_{SG}>|v_{TP}|}_{v_{SD}<v_{SGT}} = v_{SD}\left(\frac{W}{L}\right) K_P'\left(v_{SG} - |v_{TP}| - \frac{v_{SD}}{2}\right)$$

$$|v_{TP}| = |V_{TP0}| + \gamma_P\left(\sqrt{2\psi_B - v_{SB}} - \sqrt{2\psi_B}\right)$$

$$v_{SD(SAT)} = v_{SG} - |v_{TP}| \equiv v_{SGT}$$

$$i_D\Big|^{v_{SG}>|v_{TP}|}_{v_{SD}>v_{SGT}} \approx \left(\frac{W}{L}\right)\left(\frac{K_P'}{2}\right)(v_{SG} - |v_{TP}|)^2(1 + \lambda_P v_{SD})$$

$$\approx \sqrt{\frac{2i_D}{(W/L)K_P'(1 + \lambda_P v_{SD})}}$$

Weak Inversion: $\quad v_{GS} \approx v_T \pm 50$ mV

Conduction: $\quad$ Carriers diffuse & drift

Small-Signal Transconductance:

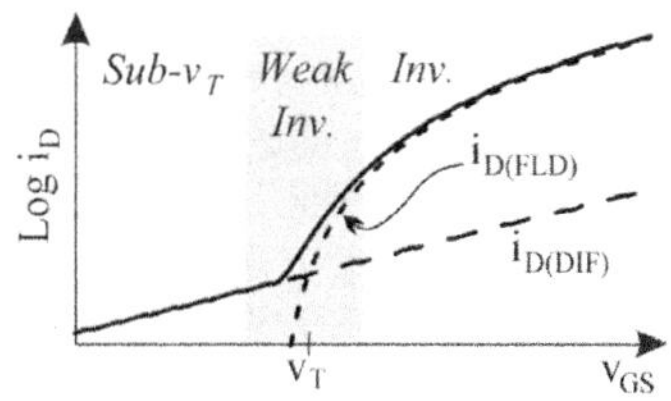

$$g_m\Big|^{0<v_{GS}<v_T}_{v_{DS}>3V_t} \equiv \frac{\partial i_D}{\partial v_{GS}} \approx \left(\frac{1}{n_I V_t}\right)\left(\frac{W}{L}\right) I_S \exp\left(\frac{v_{GS} - v_T}{n_I V_t}\right) \approx \frac{i_D}{n_I V_t}$$

$$v_{DS(SAT)} = 2n_I V_t$$

↓

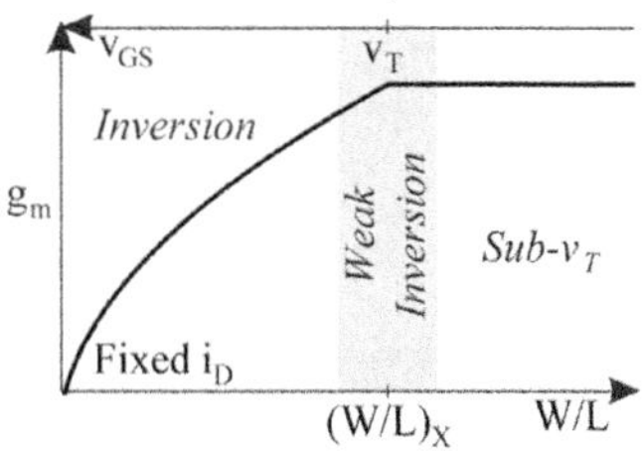

$$g_m\Big|^{v_{GS}>v_T}_{v_{DS}>v_{GST}} \equiv \frac{\partial i_D}{\partial v_{GS}} \approx \left(\frac{W}{L}\right) K'(v_{GS} - v_T)(1 + \lambda v_{DS})$$

$$= \sqrt{2i_D K'(W/L)(1 + \lambda v_{DS})}$$

$$g_{m(MAX)}\Big|^{v_{GS}>v_T}_{v_{DS}>v_{GST}} \approx \sqrt{2i_D K'(W/L)_X(1 + \lambda v_{DS})} = g_m\Big|^{0<v_{GS}<v_T}_{v_{DS}>3V_t} \approx \frac{i_D}{n_I V_t}$$

$$\therefore \left(\frac{W}{L}\right)_X \approx \frac{i_D}{2n_I^2 V_t^2 K'(1 + \lambda v_{DS})} \quad \rightarrow \quad v_{DS(SAT)}\Big|^{v_{GS}>v_T}_{(W/L)_X} \approx \sqrt{\frac{2i_D}{(W/L)_X K'(1 + \lambda v_{DS})}} = 2n_I V_t$$

C. Capacitances

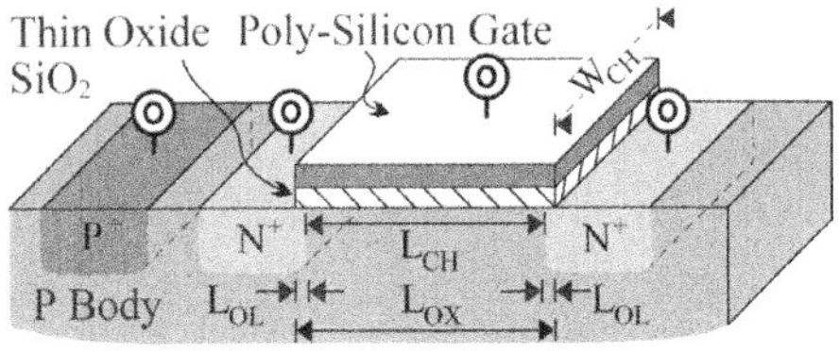

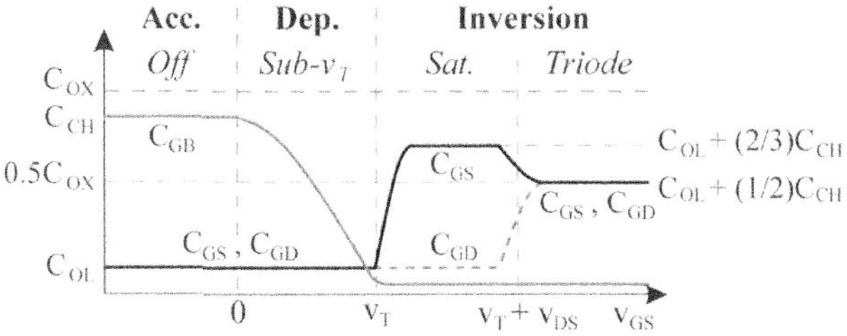

PN Junctions: C_{SB} C_{DB}

$$C_J \approx C_{DEP} = \frac{A_J C_{J0}''}{\sqrt{1 + \frac{V_{JR}}{V_{BI}}}}$$

Gate Oxide: C_{GS} C_{GB} C_{GD}

$$L_{OX(MIN)} \equiv L_{MIN}$$

Cut Off

$$C_{OL} = C_{OX}''W_{CH}L_{OL} \rightarrow L_{OL} \approx \frac{L_{OX(MIN)}}{6}$$

Sub-v_T

Triode Inversion

$$C_{CH} = C_{OX}''W_{CH}L_{CH} \rightarrow L_{CH} = L_{OX} - 2L_{OL}$$

Saturated Inversion

$$C_{GB} = C_{CH} \oplus C_{DEP} \le \text{Min}\{C_{CH}, C_{DEP}\}$$

Example: Determine C_{GS} & C_{GD} in saturated inversion when W = 10 μm, $L = L_{MIN}$ = 180 nm, L_{OL} = 30 nm, C_{OX}'' = 2.76 fF/μm².

Solution:

$$C_{OL} = C_{OX}''W_{CH}L_{OL} = (2.76\text{m})(10\mu)(30\text{n}) = 0.83 \text{ fF}$$

$$C_{CH} = C_{OX}''W_{CH}L_{CH} = C_{OX}''W_{CH}\left(L_{OX} - 2L_{OL}\right)$$

$$= (2.76\text{m})(10\mu)[180\text{n} - 2(30\text{n})] = 3.3 \text{ fF}$$

$$C_{GS} = C_{OL} + (2/3)C_{CH} = 0.83\text{f} + (2/3)(3.3\text{f}) = 3.0 \text{ fF}$$

$$C_{GD} = C_{OL} = 0.83 \text{ fF}$$

D. MOS Diodes

Diode Connection

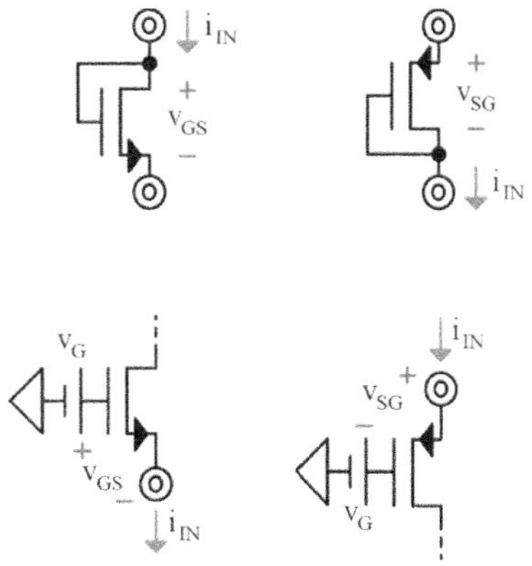

Wired: i_{IN} charges C_{GS} until i_{DS} pulls i_{IN}.

i_{IN} may or may not invert a channel.

Sub-v_T → $v_{GS} \approx v_T + n_I V_t \ln\left[\frac{i_{IN}}{(W/L)I_{SN}}\right] \le v_T$

Inverted → $v_{DS(SAT)} = v_{GS} - v_T$

$$v_{GS} = v_T + v_{DS(SAT)} = V_{T0} + \gamma\left(\sqrt{2\psi_B - v_{BS}} - \sqrt{2\psi_B}\right) + \sqrt{\frac{2i_D}{(W/L)K'(1+\lambda v_{DS})}}$$

Body effect (γ term) can be significant.

L_{CH} modulation (λ term) can be negligible, but less negligible with short L.

Implicit Diode Action: Induced by circuit

Example: Determine v_{GS} when i_{IN} pulls 100 μA from v_S & $v_G = v_D = v_B = 0$ V,

W = 10 μm, L = 180 nm, L_{OL} = 30 nm, n_I = 1.75, V_{TN0} = 400 mV,

K_N' = 200 μA/V², γ_N = 600 m√V, ψ_B = 300 mV, λ_N = 10%.

Solution: $L_{CH} = L - 2L_{OL} = 180n - 2(30n) = 120$ nm

$$v_{DS(SAT)} = \sqrt{\frac{2i_{IN}}{(W/L)_{CH}K_N'(1+\lambda_N v_{DS})}} \approx \sqrt{\frac{2i_{IN}}{(W/L)_{CH}K_N'}} = \sqrt{\frac{2(100\mu)}{(10\mu/120n)(200\mu)}}$$

$$= 110 \text{ mV} > 2n_I V_t = 2(1.75)(25.6m) = 90 \text{ mV} \quad \therefore \quad \text{Inverted}$$

$v_{GS} = v_{DS} = v_{BS} = 340$ mV $< V_{TN0}$ → v_{BS} reduces v_{TN} more than $V_{DS(SAT)}$ raises v_{GS}

$$\approx v_{TN} + v_{DS(SAT)} = V_{TN0} + \gamma_N\left(\sqrt{2\psi_B - v_{BS}} - \sqrt{2\psi_B}\right) + \sqrt{\frac{2i_{IN}}{(W/L)_{CH}K_N'(1+\lambda_N v_{DS})}}$$

$$= 400m + 600m\left(\sqrt{2(300m) - v_{GS}} - \sqrt{2(300m)}\right) + \sqrt{\frac{2(100\mu)}{(10\mu/120n)(200\mu)(1+10\% v_{GS})}}$$

E. Junction Isolation

Integration: Shared substrate

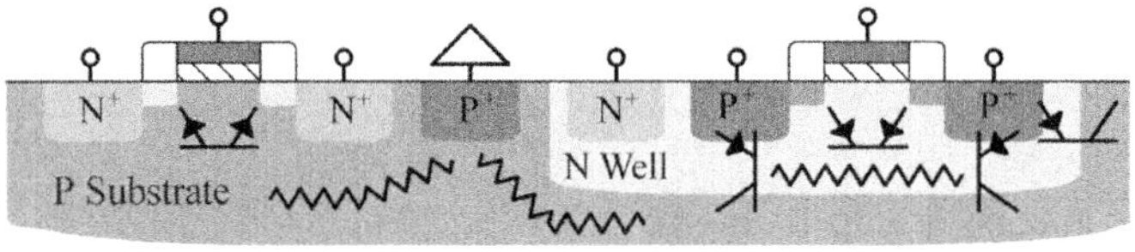

Parasitic Components:

Substrate Diodes: D_{BS} D_{BD} Channel BJTs: Q_{CH} Substrate BJTs: Q_{SB} Q_{DB}

Deactivate Substrate Devices: Reverse-bias PN junctions

Diodes: P substrate to low voltage BJTs: N bodies to high voltage

MOSFETs: Substrate Welled Approximations:

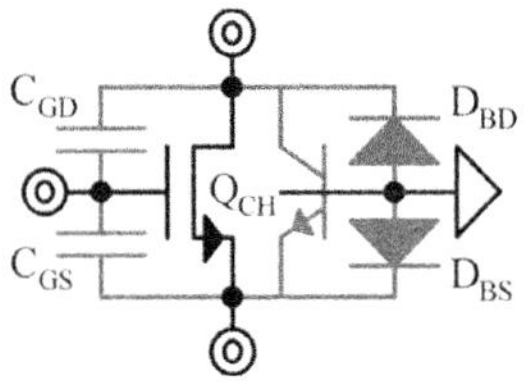

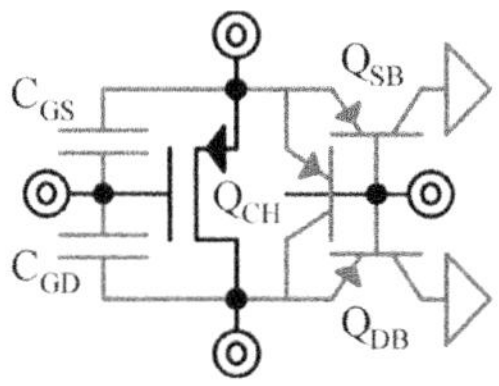

$Q_{CH} \approx$ Negligible

$Q_{S/DB} \rightarrow D_{S/DB}$

Chapter 3. Switched Inductors

3.1. Transfer Media

3.2. Switched Inductor

3.3. Buck–Boost

3.4. Buck

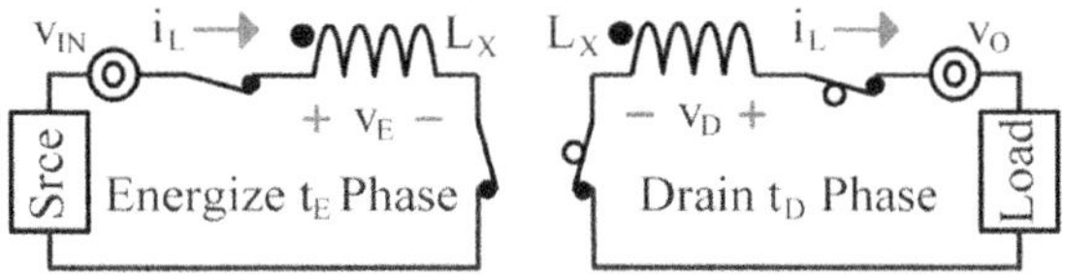

3.5. Boost

3.6. Flyback

3.1. Transfer Media: A. Inductor

$$v_L = L_X\left(\frac{di_L}{dt_X}\right) = L_X\left(\frac{\Delta i_L}{\Delta t_X}\right)$$

Ideal Inductor

$$E_M \equiv E_L = \int P_L dt_X = \int i_L v_L dt_X = \left(\frac{1}{2}\right) L_X i_L^{\ 2}$$

Source v_L magnetizes/energizes L_X

Reverse v_L demagnetizes/drains L_X

Applied ↓

$$i_L = \left(\frac{v_L}{L_X}\right) t_X$$

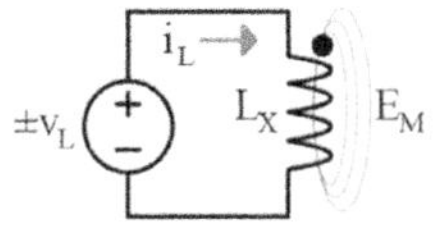

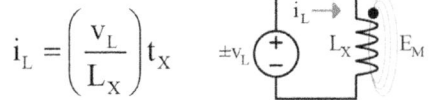

Actual Inductor

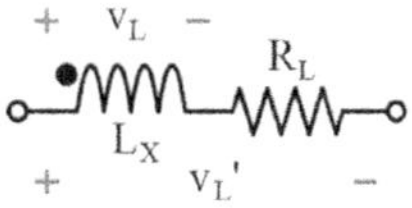

$L_X \propto$ Loop's cross-sectional area A_L $\quad$ $R_L \propto 1 / A_C$

$\propto$ Number of turns N_L $\quad$ $\propto N_L$

Proximity Effect: Nearby fields push i_L away from edges → N_L raises R_L

Skin Effect: Fast-moving charges flow through outer rim → $R_{L(AC)} > R_{L(DC)}$

Magnetic Resistance: Molecules impede dipole alignment → $R_{L(MAG)}$

B. Transformer

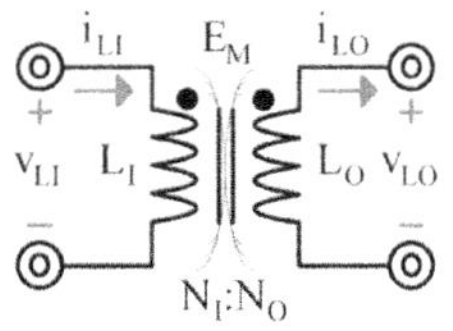

Ideal Transformer: $k_C = 1$

Coupled L_I & L_O share magnetic space

$\therefore \quad E_M = 0.5L_I i_{LI}^2 = 0.5L_O i_{LO}^2 \quad \rightarrow \quad \dfrac{i_{LI}}{i_{LO}} = \sqrt{\dfrac{L_O}{L_I}} \equiv k_L$ ← Translation

v_{LI} energizes L_I & v_{LO} drains L_O — If geometries match: $k_L = \dfrac{N_O}{N_I}$

$\therefore \quad P_{IN} = i_{LI}v_{LI} = P_O = i_{LO}v_{LO} \quad \rightarrow \quad \dfrac{v_{LO}}{v_{LI}} = \dfrac{i_{LI}}{i_{LO}} = k_L$

Actual Transformer: Coupling Factor $\equiv k_C < 1$

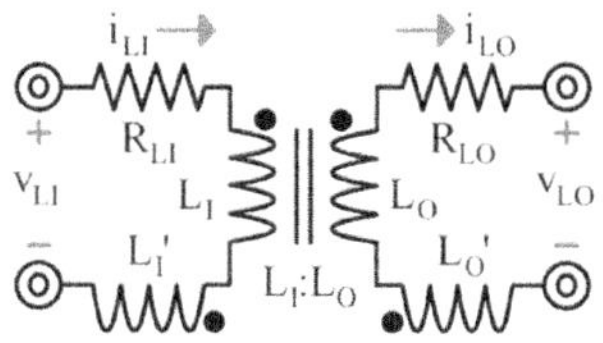

Separation & misalignment reduce k_L'

→ Uncoupled L_I' & L_O'

$\therefore \quad L_O$ avails k_{CI} fraction of E_{IN} $\rightarrow \quad k_L' \equiv \dfrac{v_{LO}}{v_{LI}} = \dfrac{v_{LI}k_{CI}k_L}{v_{LI}} = \left(\dfrac{L_I}{L_I + L_I'}\right)\sqrt{\dfrac{L_O}{L_I}}$

3.2. Switched Inductors: A. Inductor Current

Operation:

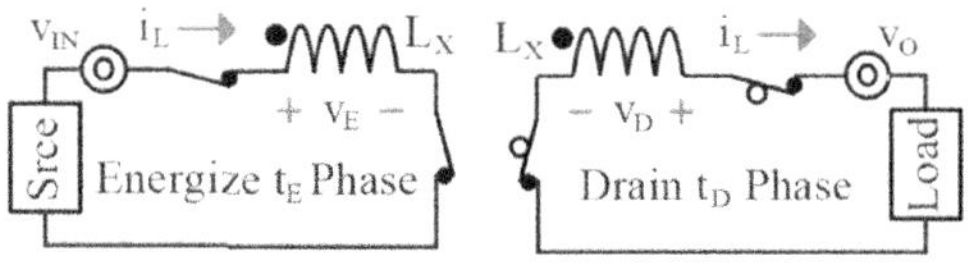

$$\frac{di_L}{dt_X} = \frac{v_L}{L_X}$$

v_{IN} energizes L_X $\therefore \quad v_E = v_{IN}$

Reverse v_O drains L_X $\therefore \quad v_D = v_O$

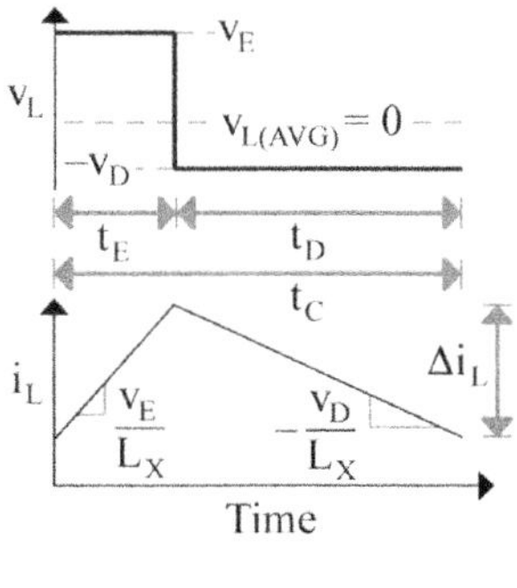

$d_E \equiv \dfrac{t_E}{t_C} \qquad d_D \equiv \dfrac{t_D}{t_C} = \dfrac{t_C - t_E}{t_C} = 1 - d_E$

Static steady-state operation:

$$v_{L(AVG)} = v_E d_E - v_D d_D = 0 \quad \therefore \quad \frac{v_E}{v_D} = \frac{d_D}{d_E} \quad \rightarrow \quad d_E = \frac{v_D}{v_E + v_D}$$

$i_{L(LO)}$ & $i_{L(HI)}$ do not vary $\therefore \quad \Delta i_L = \left(\dfrac{v_E}{L_X}\right)t_E = \left(\dfrac{v_D}{L_X}\right)t_D$ → Equal V·s

B. Continuous Conduction

$$\frac{di_L}{dt_X} \neq 0: \quad d_E\big|_{CCM} = \frac{t_E}{t_C}\bigg|_{CCM} = \frac{t_E}{t_{SW}}$$

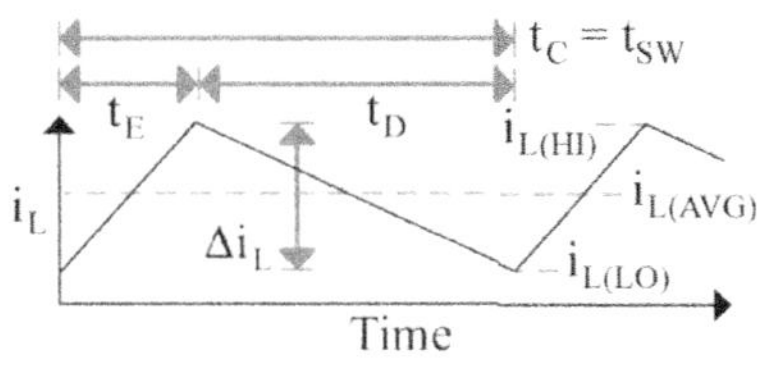

$$d_D\big|_{CCM} = \frac{t_D}{t_C}\bigg|_{CCM} = \frac{t_D}{t_{SW}} = 1 - d_E$$

$$i_{L(AVG)}\big|_{CCM} = i_{L(LO)} + \Delta i_{L(AVG)}\big|_{CCM} = i_{L(LO)} + 0.5\Delta i_L$$

C. Discontinuous Conduction

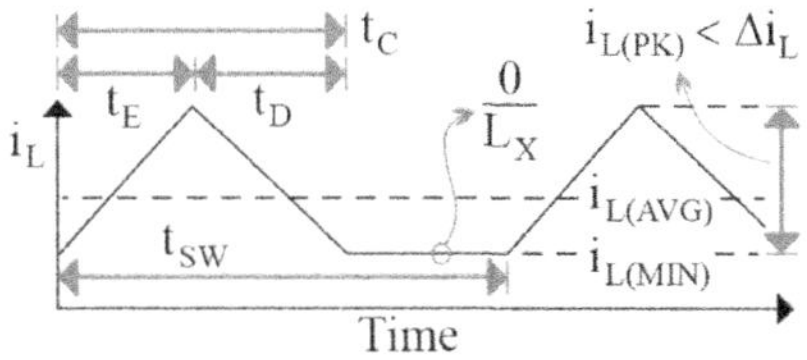

$$\frac{di_L}{dt_X} >=< 0: d_E\big|_{DCM} = \frac{t_E}{t_C} \neq \frac{t_E}{t_{SW}}$$

DCM: $i_{L(MIN)} = 0$

Pseudo DCM: $i_{L(MIN)} > 0$

$$d_D\big|_{DCM} = \frac{t_D}{t_C} = 1 - d_E \neq \frac{t_D}{t_{SW}}$$

$$i_{L(AVG)}\big|_{P/DCM} = i_{L(MIN)} + \Delta i_{L(AVG)}\big|_{P/DCM} = i_{L(MIN)} + \left(\frac{i_{L(PK)}}{2}\right)\left(\frac{t_C}{t_{SW}}\right) < i_{L(MIN)} + \frac{\Delta i_L}{2}$$

Example: Determine t_E, t_D, & Δi_L in CCM when

$$v_E = 2\ \text{V},\ v_D = 1\ \text{V},\ t_{SW} = 1\ \mu\text{s},\ L_X = 10\ \mu\text{H}.$$

Solution:

$$d_E = \frac{v_D}{v_E + v_D} = \frac{1}{2+1} = 33\%$$

$$\therefore \quad t_E = d_E t_C = d_E t_{SW} = (33\%)(1\mu) = 330\ \text{ns}$$

$$t_D = t_C - t_E = t_{SW} - t_E = 1\mu - 330\text{n} = 670\ \text{ns}$$

$$\Delta i_L = \left(\frac{v_E}{L_X}\right)t_E = \left(\frac{2}{10\mu}\right)(330\text{n}) = 66\ \text{mA}$$

Example: Determine t_E, t_D, & Δi_L when $v_E = 2$ V, $v_D = 1$ V, $i_{L(AVG)} = 25$ mA,

i_L valleys to 0 A, $t_{SW} = 1$ µs, $L_X = 10$ µH.

Solution: $d_E = 33\%$, $\Delta i_L = 66$ mA in CCM from previous example

$$i_{L(AVG)} = 25\ \text{mA} < i_{L(LO)} + \left.\frac{\Delta i_L}{2}\right|_{CCM} = 0 + \frac{66m}{2} = 33\ \text{mA} \quad \therefore \quad \text{DCM}$$

$$i_{L(AVG)} = \left(\frac{i_{L(PK)}}{2}\right)\left(\frac{t_C}{t_{SW}}\right) \qquad\qquad i_{L(PK)} = \left(\frac{v_E}{L_X}\right)t_E = \left(\frac{v_E}{L_X}\right)d_E t_C$$

$$t_C = \frac{2i_{L(AVG)}t_{SW}}{i_{L(PK)}} = \sqrt{2i_{L(AVG)}\left(\frac{L_X}{v_E}\right)\left(\frac{t_{SW}}{d_E}\right)} = \sqrt{2(25m)\left(\frac{10\mu}{2}\right)\left(\frac{1\mu}{33\%}\right)} = 870\ \text{ns}$$

$$t_E = d_E t_C = (33\%)(870n) = 290\ \text{ns} \qquad\qquad t_D = t_C - t_E = 870n - 290n = 580\ \text{ns}$$

$$i_{L(PK)} = \left(\frac{v_E}{L_X}\right)t_E = \left(\frac{2}{10\mu}\right)(290n) = 58\ \text{mA} \ < \ \Delta i_L = 66\ \text{mA in CCM}$$

D. Ohmic Loss

$v_R = i_L R_{ESR} \propto i_L$

→ Reduces v_E ∴ Longer t_E' needed

Raises v_D ∴ Shortens t_D'

→ Burns $E_R \propto i_L^2$ ∴ Longer t_E' needed

Shortens t_D'

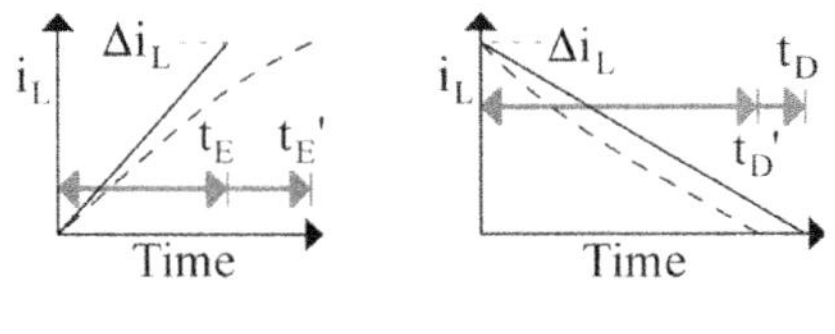

Nonlinear i_L Higher d_E' for same v_O

$$d_E' = \frac{v_D}{v_E + v_D}$$

→ Ideal v_D When $R_E \approx R_D$

$$= \frac{v_{DI} + v_{RL} + v_{RD}}{(v_{EI} - v_{RL} - v_{RE}) + (v_{DI} + v_{RL} + v_{RD})} = \frac{v_{DI} + v_{RL} + v_{RD}}{(v_{EI} - v_{RE}) + (v_{DI} + v_{RD})} \approx \frac{v_{DI} + v_{RL} + v_{RD}}{v_{EI} + v_{DI}}$$

→ Ideal v_E ∴ $d_E' > d_E$ v_B or controller sets v_O

Note: Open-loop simulations often fix d_E ∴ v_O shifts with v_R's

E. CMOS Implementation

Channel Resistance: Sheet Resistance → Ohms per Square

$$R_{CH}\Big|_{v_{DS} << V_{DS(SAT)}}^{|v_{GS}| > v_T} = \left(\frac{L}{W}\right) R_{SH} \approx \left(\frac{L}{W}\right)\left(\frac{1}{K_{N/P}'|v_{GST}|}\right) = \left(\frac{L}{W}\right)\left(\frac{1}{\mu_{N/P} C_{OX}''|v_{GST}|}\right)$$

(↑ R_{SH}: Sheet Resistance; ↓ $\frac{L}{W}$: # of Squares)

MOS Type: # of Squares

Select highest $K_{N/P}'|v_{GST}|$ → Select PMOS when $K_P'v_{SGT} \geq K_N'v_{GST}$

∴ Typically, NMOS for low v_S PMOS for high v_S

Body Connection:

Block undesired i_{IN} → Short body diode that conducts i_{IN}

Block reverse i_O → Short body diode that conducts negative i_O

F. Design Limits

Losses: $i_L \approx$ Linear when $P_R \approx$ Low

$\therefore\ v_{R(MAX)} = i_{L(HI)} R_{SER} << v_{L(MIN)}$

Predictable i_L: $i_L \approx$ Linear $\therefore\ \Delta v_{O(MAX)} << v_O$ → Add C_O

Oscillations: L_X & C_O exchange energy every $0.25t_{LC}$

→ $t_{LC} = 2\pi\sqrt{L_X C_O}$ $\therefore\ t_{SW} << 25\% t_{LC}$

Example: 10% limits for R_{CH} & f_{SW} when two switches conduct i_L, $v_D = 1$ V, $v_E = 2$ V, $L_X = 10$ μH, $R_L = 200$ mΩ, $C_O = 10$ μF, $i_{L(AVG)} = 70$ mA, $\Delta i_L = 30$ mA.

Solution:

$$R_{SER} < \frac{10\% v_{L(MIN)}}{i_{L(HI)}} = \frac{10\% v_D}{i_{L(AVG)} + 0.5\Delta i_L} = \frac{10\%(1)}{70m + 0.5(30m)} = 1.2\ \Omega$$

$$\rightarrow\ R_{CH} < \frac{R_{SER} - R_L}{2} = \frac{1.2 - 200m}{2} = 500\ m\Omega$$

$$t_{SW} < 2.5\%(2\pi)\sqrt{L_X C_O} = 1.6\ \mu s\ \rightarrow\ f_{SW} \geq 620\ kHz$$

3.3. Buck–Boost: A. Ideal

Operation: S_{EI}:S_{EG} energize L_X from v_{IN}

S_{DG}:S_{DO} drain L_X into v_O

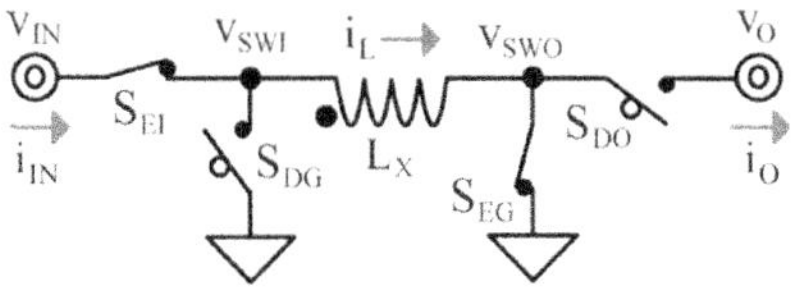

L_X = Low-f_O short → $v_{SWI(AVG)} = v_{SWO(AVG)}$

If $d_E < 50\%$, $d_D > 50\%$ ∴ $0.5v_{IN} > v_{SW(AVG)}$'s $> 0.5v_O$ → Buck v_{IN} to v_O

If $d_E > 50\%$, $d_D < 50\%$ ∴ $0.5v_{IN} < v_{SW(AVG)}$'s $< 0.5v_O$ → Boost v_{IN} to v_O

$$\rightarrow\ v_{SW(AVG)} = v_{SWI(AVG)} = v_E\left(\frac{t_E}{t_C}\right) = v_{IN}d_E = v_{SWO(AVG)} = v_D\left(\frac{t_D}{t_C}\right) = v_O d_D = v_O\left(1-d_E\right)$$

$$\therefore\ v_O = v_{IN}\left(\frac{d_E}{d_D}\right) = v_{IN}\left(\frac{d_E}{1-d_E}\right) \ \rightarrow\ d_E = \frac{v_D}{v_E + v_D} = \frac{v_O}{v_{IN} + v_O}$$

$$E_{IN} \overset{\text{Ideal}}{=} E_L \underset{\text{Ideal}}{=} E_O$$

$$P_O = v_O i_{O(AVG)} = v_O i_{L(AVG)} d_D \qquad P_{IN} = v_{IN} i_{IN(AVG)} \approx v_{IN} i_{L(AVG)} d_E = v_{IN}\left(\frac{i_{O(AVG)}}{d_D}\right)d_E \overset{\text{Ideal}}{=} P_O$$

B. Asynchronous

Concept: Once energized, stored i_L can close drain diodes automatically.

Drain Diodes: S_{DG} → D_{DG} S_{DO} → D_{DO}

Energize FETs: S_{EG} → Low v_{SW} → NMOS

S_{EI} → High v_{SW} → PMOS

Body diodes should block i_{IN}

∴ v_{IN} to $v_{EI(B)}$ & $v_{EG(B)}$ to Gnd

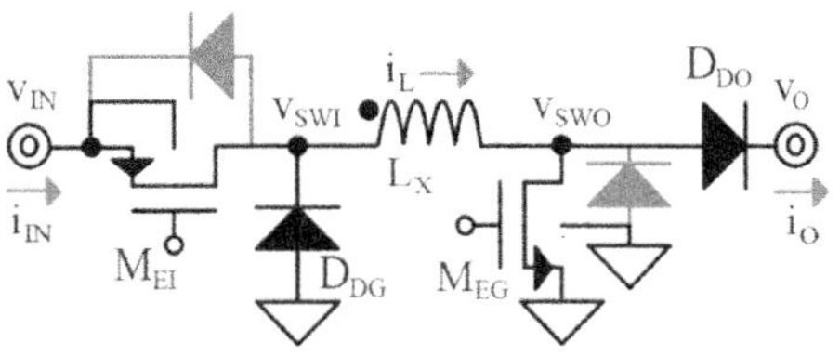

$v_{SWI(AVG)} \approx v_{IN}d_E' - v_{DB}d_D' = v_{IN}d_E' - v_{DB}(1 - d_E')$

$v_{SWO(AVG)} \approx (0)d_E' + (v_O + v_{DO})d_D' = (v_O + v_{DO})(1 - d_E')$

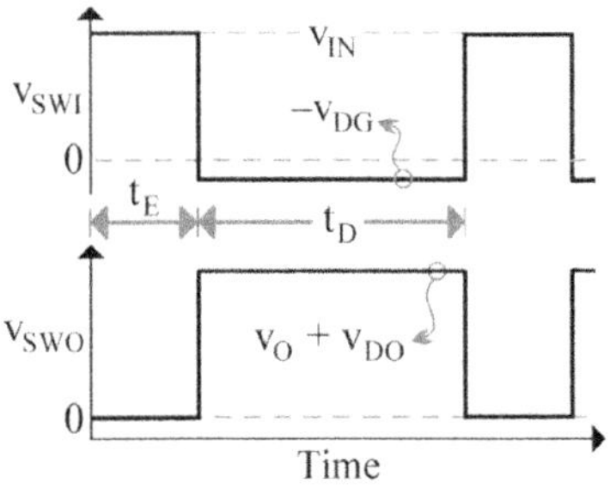

$$\frac{d_E'}{d_D'} = \frac{v_D}{v_E} \approx \frac{v_O + v_{DG} + v_{DO}}{v_{IN}} = \frac{v_O}{v_{IN}} + \frac{v_{DG} + v_{DO}}{v_{IN}} \qquad d_E' = \frac{v_D}{v_E + v_D} \approx \frac{v_O + v_{DO} + v_{DG}}{v_{IN} + v_O + v_{DG} + v_{DO}} > d_E$$

SPICE Simulations

Structure:

ASCII Text File:	[name].cir
First (Title) Line:	[text]
Comment Lines:	* [text]
Net List:	[circuit]
Model Lines:	.model [model definition]
Command Lines:	.[command]
Last Line (End of File):	.end

Switches:

Switch:	S[name] node1 node2 vnode+ vnode– [model]
Diode:	D[name] anode cathode [model] [area multiplier]
MOSFET:	M[name] d g s b [model] W=[width] L=[length] M=[multiplier]

```
* Ideal Buck-Boost in CCM
vde de 0 dc=0 pulse 0 1 50n 1n 1n 670n 1u
vin vin 0 dc=2
sei vin vswi de 0 sw1v
ddg 0 vswi idiode
lx vswi vswo 10u
seg vswo 0 de 0 sw1v
ddo vswo vo idiode
co vo 0 5u
ro vo 0 10
.model sw1v vswitch roff=1e12 ron=1m voff=450m von=550m
.model idiode d is=1f n=0.001
.tran 700u
.end
```

(n=0.001 ↓ Ideality factor)

Stimulus:

Pulse v1 v2 tdelay trise tfall v2width tper

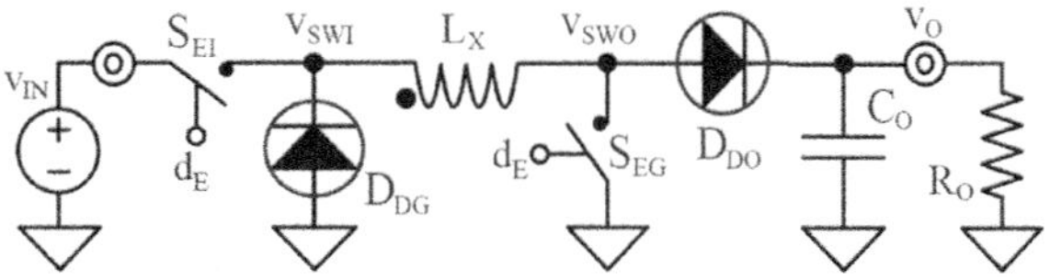

* Plot v_O & i_L across 700 μs

* Plot v_O, i_L, v_{SWI}, v_{SWO} across last 5 μs

* Asynchronous Buck-Boost in CCM

```
vge vge 0 dc=0 pulse 0 2 50n 1n 1n 740n 1u
vgeb vgeb 0 dc=2 pulse 2 0 50n 1n 1n 740n 1u
vin vin 0 dc=2
mei vswi vgeb vin vin pmos0 w=300m l=250n
ddg 0 vswi diode1
lx vswi vswo 10u
ddo vswo vo diode1
meg vswo vge 0 0 nmos0 w=100m l=250n
co vo 0 5u
ro vo 0 10
.model nmos0 nmos vto=1 kp=200u lambda=0.1 tox=5n ld=30n cgso=200p cgdo=200p
.model pmos0 pmos vto=-1 kp=40u lambda=0.1 tox=5n ld=30n cgso=200p cgdo=200p
.model diode1 d is=50f n=1 tt=1n
.tran 700u
.end
```

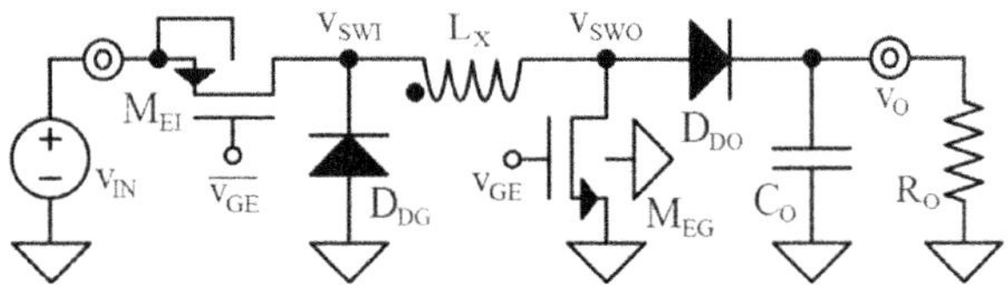

Overlap C_{OX} per $W_{CH} = C_{OX}''L_{OL}$

L_{OL}

Conduction Modes:

Continuous

Discontinuous

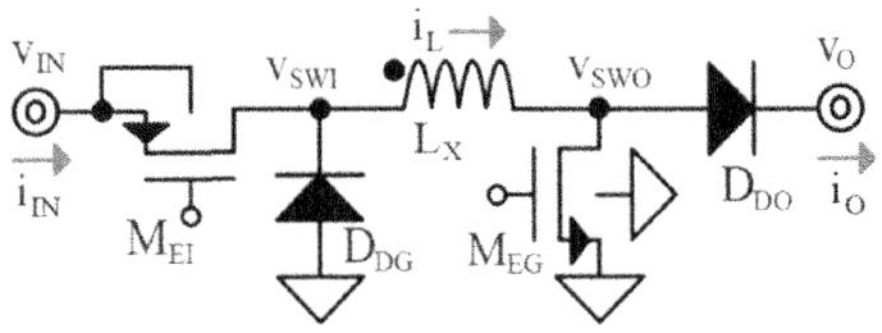

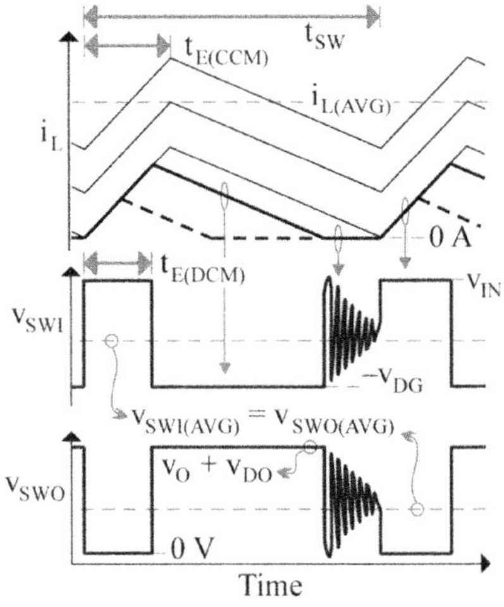

Off Period:

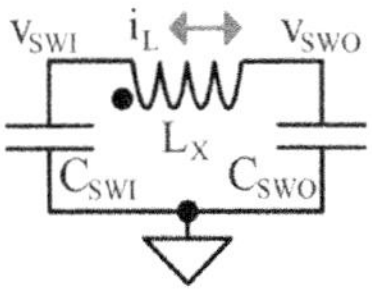

After t_C: $v_{SWO} \approx v_O + v_{DO}$ → E_{CSWO}

$v_{SWI} \approx -v_{DG}$ → E_{CSWI}

$v_L = v_{SWO} - v_{SWI} = v_O + 2v_D$

∴ C_{SW}'s & L_X exchange energy until R_{SER} burns it → Damped oscillations

C. Synchronous

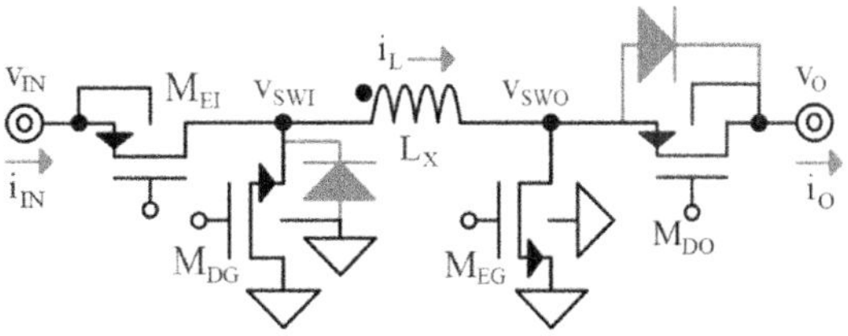

Energize FETs: $S_{EI} \rightarrow M_{EI}$ $\quad S_{EG} \rightarrow M_{EG}$

Drain FETs: $\quad S_{DG} \rightarrow$ Low $v_{SW} \rightarrow$ NMOS

$S_{DO} \rightarrow$ High $v_{SW} \rightarrow$ PMOS

Body diodes should block reverse i_O

$\therefore$ v_O to $v_{DO(B)}$ & $v_{DG(B)}$ to Gnd

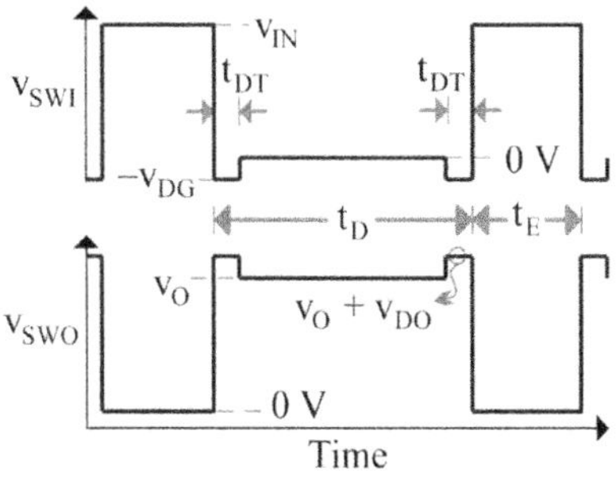

Caution: FETs can short v_{IN} & v_O to Gnd

→ Insert dead time t_{DT} $\therefore$ D_D's steer $i_{L(DT)}$

$$v_{SWI(AVG)} \approx v_{IN} d_E'' - v_{DG}\left(\frac{2t_{DT}}{t_{SW}}\right) + (0)\left(\frac{t_D - 2t_{DT}}{t_{SW}}\right)$$

$$v_{SWO(AVG)} \approx (0) d_E'' + (v_O + v_{DO})\left(\frac{2t_{DT}}{t_{SW}}\right) + v_O\left(\frac{t_D - 2t_{DT}}{t_{SW}}\right) = v_O(1 - d_E'') + v_{DO}\left(\frac{2t_{DT}}{t_{SW}}\right)$$

$$d_E'' = \frac{v_D}{v_E + v_D} \approx \frac{v_O}{v_{IN} + v_O} + \left(\frac{v_{DO} + v_{DG}}{v_{IN} + v_O}\right)\left(\frac{2t_{DT}}{t_{SW}}\right) = d_E + \left(\frac{v_{DO} + v_{DG}}{v_{IN} + v_O}\right)\left(\frac{2t_{DT}}{t_{SW}}\right) > d_E$$

Example: Determine ideal d_E when $v_{IN} = 2$ V, $v_O = 1$ V.

Solution: $d_E = \frac{v_D}{v_E + v_D} = \frac{v_O}{v_{IN} + v_O} = \frac{1}{2+1} = 33\% < 50\%$ → Buck

Example: Ideal d_E when $v_{IN} = 2$ V, $v_O = 4$ V.

Solution: $d_E = \frac{v_D}{v_E + v_D} = \frac{v_O}{v_{IN} + v_O} = \frac{4}{2+4} = 67\% > 50\%$ → Boost

Example: Asynchronous boost d_E' when D's drop 800 mV.

Solution: $d_E' = \frac{v_D}{v_E + v_D} \approx \frac{v_O + v_{DO} + v_{DG}}{v_{IN} + v_O + v_{DG} + v_{DO}} = \frac{4 + 800m + 800m}{2 + 4 + 800m + 800m} = 74\% > d_E$

Reason: D_{DG} & D_{DO} raise v_D → Shorten t_D & t_C $\therefore$ t_E = Higher t_C fraction

Example: Synchronous boost d_E'' when D's drop 800 mV, $t_{DT} = 50$ ns, $t_{SW} = 1$ μs.

Solution: $d_E'' \approx \frac{v_O}{v_{IN} + v_O} + \left(\frac{v_{DO} + v_{DG}}{v_{IN} + v_O}\right)\left(\frac{2t_{DT}}{t_{SW}}\right) = \frac{4}{2+4} + \left(\frac{0.8 + 0.8}{2+4}\right)\left[\frac{2(50n)}{1\mu}\right] = 69\%$

Note: D_{DG} & D_{DO} raise v_D, but not as much as asynchronous diodes. $\quad > d_E$

Conduction Modes:

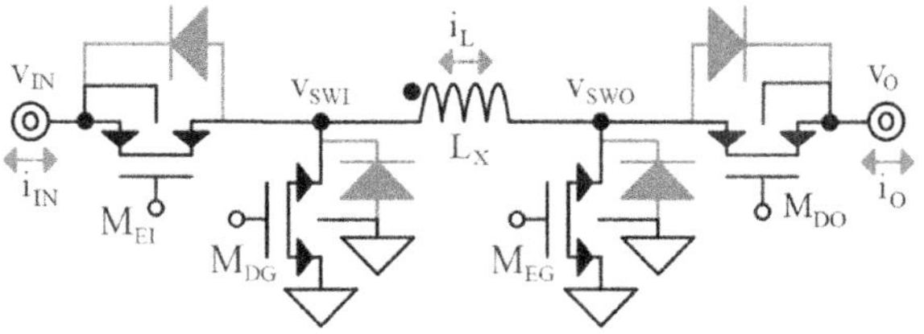

FETs can conduct reverse $i_L < 0$

∴ No DCM

D_E's conduct $i_L < 0$ across t_{DT}

Reverse i_L reverses P_O → Not good

∴ Open M_D's when $i_L = 0$ → Force DCM

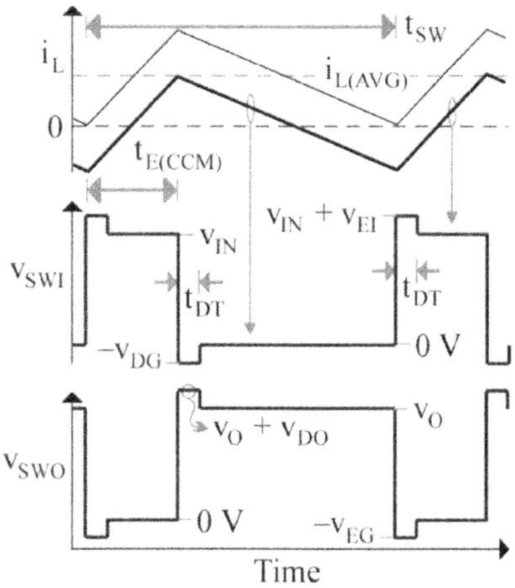

X Body diodes inject i_{SUB} when conducting i_L

√ Implicit MOS diodes steer i_L when $|v_T| < 600$ mV

√ Short D_D's with Schottky diodes

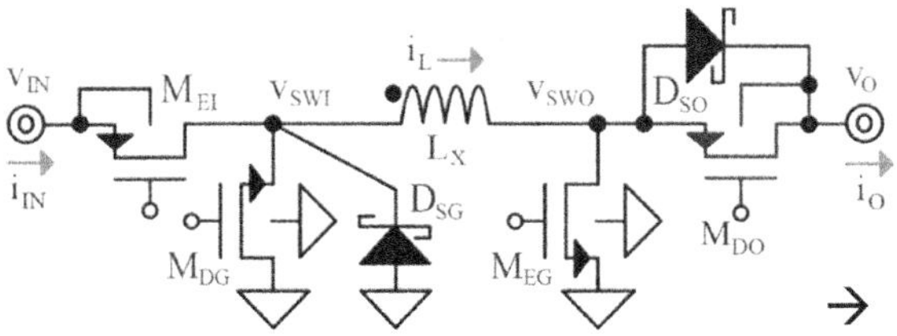

→ Lower i_{SUB}, v_D's, t_{RR}'s, but higher \$'s

3.4. Buck: A. Ideal

Concept: $v_{IN} > v_O$ ∴ v_{IN} can energize L_X into v_O → No v_O switches

Operation: S_{EI} energizes L_X from v_{IN}

S_{DG} drains L_X into v_O

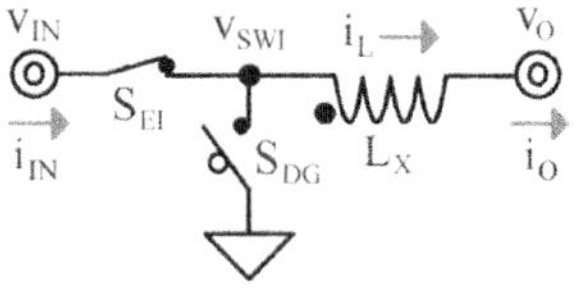

L_X = Low-f_O short: Buck v_{IN} to v_O

$$v_{SWI(AVG)} = v_{IN}\left(\frac{t_E}{t_C}\right) + (0)\left(\frac{t_D}{t_C}\right) = v_{IN}d_E = v_O < v_{IN} \quad \rightarrow \quad d_E = \frac{v_D}{v_E + v_D} = \frac{v_O}{v_{IN}}$$

$$P_O = v_O i_{O(AVG)} = v_O i_{L(AVG)} \underset{\text{Ideal}}{=} P_{IN} = v_{IN} i_{IN(AVG)} = v_{IN} i_{L(AVG)} d_E = v_{IN} i_{O(AVG)} d_E$$

Transfer: v_{IN} energizes L_X *and* supplies v_O across t_E → $E_{IN} = E_O > E_L$

Example: d_E when $v_{IN} = 2$ V, $v_O = 1$ V → $d_E = \frac{v_D}{v_E + v_D} = \frac{v_O}{v_{IN}} = \frac{1}{2} = 50\%$

B. Asynchronous

Concept: Once energized, stored i_L can close drain diode automatically.

Drain Diode: $S_{DG} \rightarrow D_{DG}$

Energize FET: $S_{EI} \rightarrow$ High $v_{SW} \rightarrow$ PMOS

Block i_{IN} $\therefore$ v_{IN} to $v_{EI(B)}$

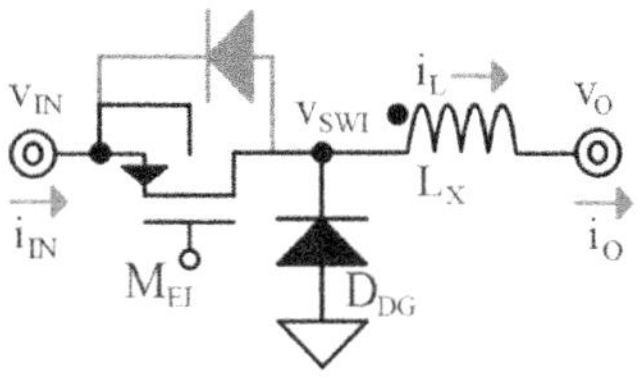

$$v_{SWI(AVG)} \approx v_{IN}d_E' - v_{DG}d_D' = v_{IN}d_E' - v_{DG}\left(1-d_E'\right) = v_O \rightarrow d_E' = \frac{v_D}{v_E + v_D} \approx \frac{v_O + v_{DG}}{v_{IN} + v_{DG}} > d_E$$

Example: d_E' when $v_{IN} = 2$ V, $v_O = 1$ V, D_{DG} drops 800 mV.

Solution: $d_E' \approx \dfrac{v_O + v_{DG}}{v_{IN} + v_{DG}} = \dfrac{1+800m}{2+800m} = 64\% > d_E = 50\%$

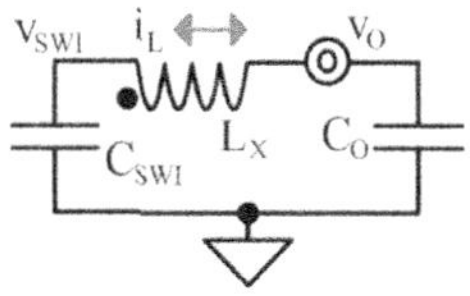

Conduction: D_{DG} blocks reverse $i_{L(D)}$ $\therefore$ DCM

$\rightarrow$ After t_C: $v_O - (-v_{DG})$ drains C's into L_X $\rightarrow$ C's & L_X exchange energy

C. Synchronous

Energize FET: $S_{EI} \rightarrow M_{EI}$

Drain FET: $S_{DG} \rightarrow$ Low $v_{SW} \rightarrow$ NMOS

Block reverse i_O $\therefore$ $v_{DG(B)}$ to Gnd

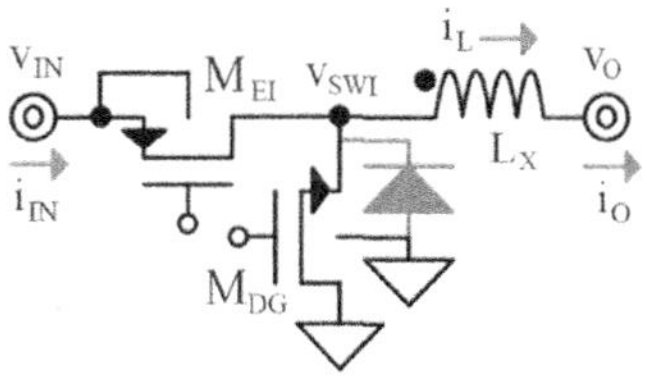

Caution: FETs can short v_{IN} to Gnd

$\rightarrow$ Insert t_{DT} $\therefore$ D_{DG} steers $i_{L(DT)}$

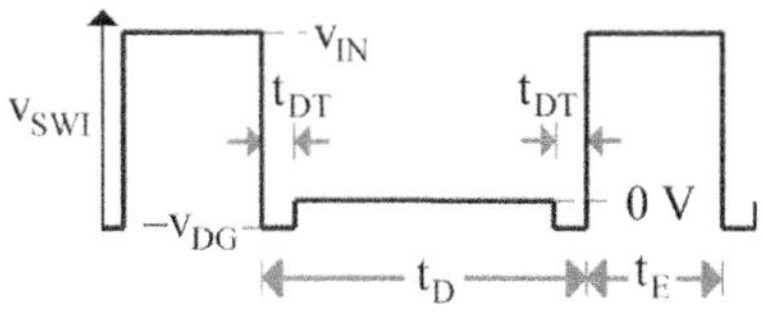

$$v_{SWI(AVG)} \approx v_{IN}d_E'' - v_{DG}\left(\frac{2t_{DT}}{t_{SW}}\right) + (0)\left(\frac{t_D - 2t_{DT}}{t_{SW}}\right) = v_O$$

$$d_E'' = \frac{v_D}{v_E + v_D} \approx \frac{v_O}{v_{IN}} + \left(\frac{v_{DG}}{v_{IN}}\right)\left(\frac{2t_{DT}}{t_{SW}}\right) = d_E + \left(\frac{v_{DG}}{v_{IN}}\right)\left(\frac{2t_{DT}}{t_{SW}}\right) > d_E$$

Example: d_E'' when $v_{IN} = 2$ V, $v_O = 1$ V, D_{DG} drops 800 mV, $t_{DT} = 50$ ns, $t_{SW} = 1$ µs.

Solution: $d_E'' \approx \dfrac{v_O}{v_{IN}} + \left(\dfrac{v_{DG}}{v_{IN}}\right)\left(\dfrac{2t_{DT}}{t_{SW}}\right) = \dfrac{1}{2} + \left(\dfrac{800m}{2}\right)\left[\dfrac{2(50n)}{1\mu}\right] = 54\% > d_E = 50\%$

3.5. Boost: A. Ideal

Concept: $v_O > v_{IN}$ ∴ v_O can drain L_X from v_{IN} → No v_{IN} switches

Operation: S_{EG} energizes L_X from v_{IN}

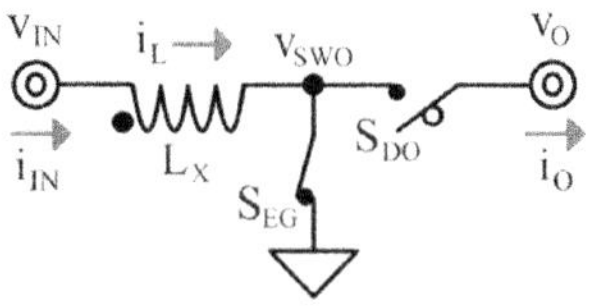

S_{DO} drains L_X into v_O

$P_O = v_O i_{O(AVG)}$

L_X = Low-f_O short: Boost v_{IN} to v_O $= v_O i_{L(AVG)} d_D$

$$v_{IN} = v_{SWO(AVG)} = (0)\left(\frac{t_E}{t_C}\right) + v_O\left(\frac{t_D}{t_C}\right) = v_O d_D = v_O(1 - d_E) < v_O \rightarrow v_O = \frac{v_{IN}}{d_D} = \frac{v_{IN}}{1-d_E}$$

$$\rightarrow d_E = \frac{v_D}{v_E + v_D} = \frac{v_O - v_{IN}}{v_O} = 1 - \frac{v_{IN}}{v_O} = 1 - d_D \qquad P_{IN} = v_{IN} i_{L(AVG)} = v_{IN}\left(\frac{i_{O(AVG)}}{d_D}\right) = P_O$$

Transfer: L_X drains into v_O across t_D *as* v_{IN} supplies v_O → $E_{IN} = E_O > E_L$ Ideal

Example: d_E when $v_{IN} = 2$ V, $v_O = 4$ V → $d_E = \frac{v_D}{v_E + v_D} = \frac{v_O - v_{IN}}{v_O} = \frac{2}{4} = 50\%$

B. Asynchronous

Concept: Once energized, stored i_L can close drain diode automatically.

Drain Diode: S_{DO} → D_{DO}

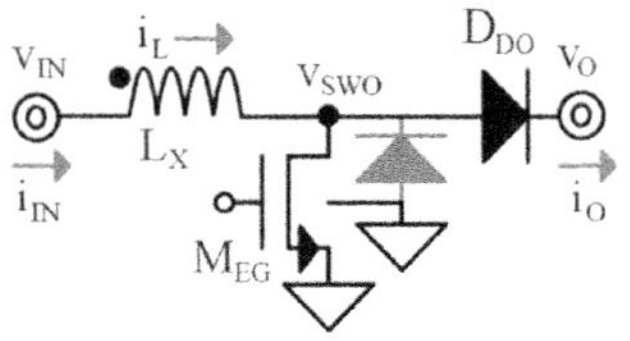

Energize FET: S_{EG} → Low v_{SW} → NMOS

Block i_{IN} ∴ $v_{EG(B)}$ to Gnd

$$v_{IN} = v_{SWO(AVG)} \approx (0)d_E' + (v_O + v_{DO})d_D' \rightarrow d_E' = \frac{v_D}{v_E + v_D} \approx \frac{v_O + v_{DO} - v_{IN}}{v_O + v_{DO}} > d_E$$

Example: d_E' when $v_{IN} = 2$ V, $v_O = 4$ V, D_{DO} drops 800 mV.

Solution: $d_E' \approx \frac{v_O + v_{DO} - v_{IN}}{v_O + v_{DO}} = \frac{4 + 800m - 2}{4 + 800m} = 58\% > d_E = 50\%$

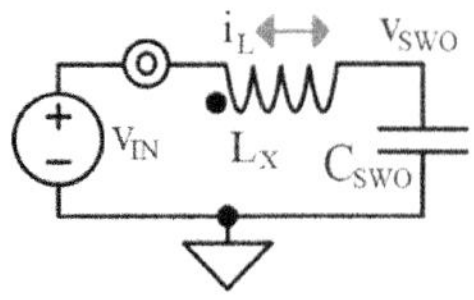

Conduction: D_{DO} blocks reverse $i_{L(D)}$ ∴ DCM

→ After t_C, $(v_O + v_{DO}) - v_{IN}$ drains C_{SWO} into L_X & v_{IN} → L & C exchange energy

C. Synchronous

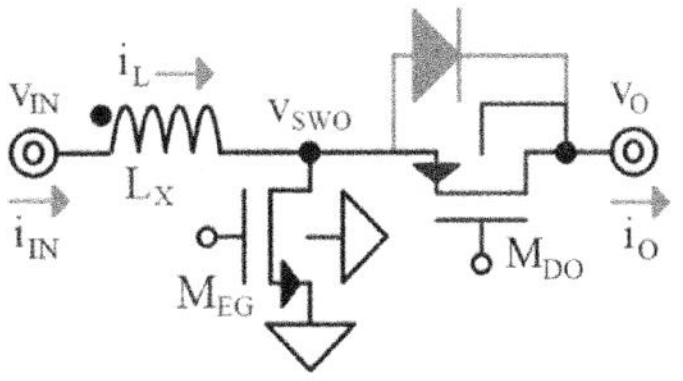

Energize FET: $S_{EG} \rightarrow M_{EG}$

Drain FET: $S_{DO} \rightarrow$ High $v_{SW} \rightarrow$ PMOS

Block reverse i_O $\therefore$ v_O to $v_{DO(B)}$

Caution: FETs can short v_O to Gnd

$\rightarrow$ Insert t_{DT} $\therefore$ D_{DO} steers $i_{L(DT)}$

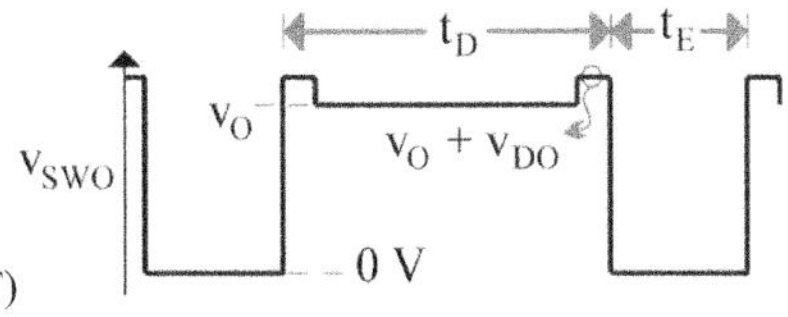

$$v_{IN} = v_{SWO(AVG)} \approx (0)d_E'' + \left(v_O + v_{DO}\right)\left(\frac{2t_{DT}}{t_{SW}}\right) + v_O\left(\frac{t_D - 2t_{DT}}{t_{SW}}\right)$$

$$d_E'' = \frac{v_D}{v_E + v_D} \approx \frac{v_O - v_{IN}}{v_O} + \left(\frac{v_{DO}}{v_O}\right)\left(\frac{2t_{DT}}{t_{SW}}\right) = d_E + \left(\frac{v_{DO}}{v_O}\right)\left(\frac{2t_{DT}}{t_{SW}}\right) > d_E$$

Example: d_E'' when $v_{IN} = 2$ V, $v_O = 4$ V, D_{DO} drops 800 mV, $t_{DT} = 50$ ns, $t_{SW} = 1$ µs.

Solution: $d_E'' \approx \frac{v_O - v_{IN}}{v_O} + \left(\frac{v_{DO}}{v_O}\right)\left(\frac{2t_{DT}}{t_{SW}}\right) = 52\% > d_E = 50\%$

3.6. Flyback: A. Ideal

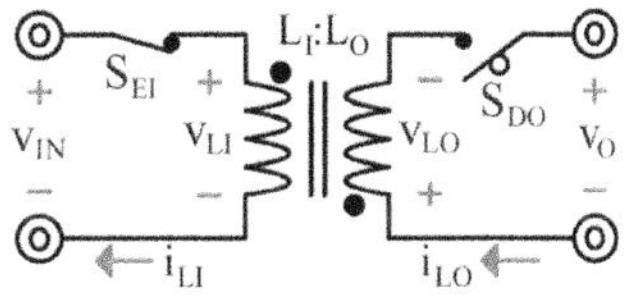

Purpose: Galvanic isolation $\rightarrow$ Separate Gnds

Operation: S_{EI} energizes L_I from v_{IN}

$\rightarrow$ v_{LO} "flies" to $v_{IN}k_L$

S_{DO} drains L_O into v_O

$\rightarrow$ v_{LI} "flies" to $-\frac{v_O}{k_L}$

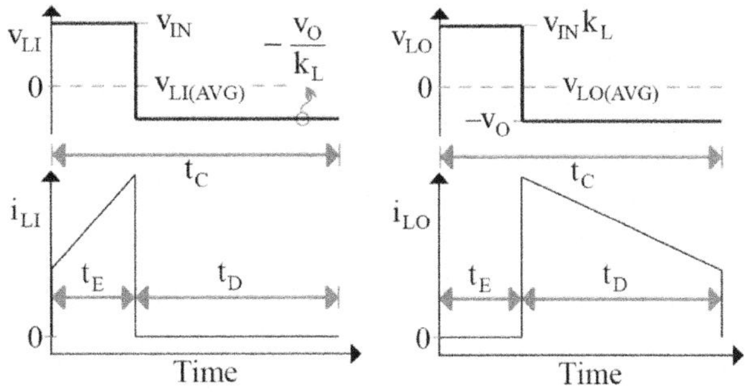

L_I & L_O = Low-f_O shorts:

$$v_{LI(AVG)} = v_{IN}d_E - \left(\frac{v_O}{k_L}\right)d_D = 0$$

$$v_{LO(AVG)} = v_{IN}k_Ld_E - v_Od_D = 0$$

$$\frac{v_O}{v_{IN}} = k_L\left(\frac{d_E}{d_D}\right) = \left(\sqrt{\frac{L_O}{L_I}}\right)\left(\frac{d_E}{1-d_E}\right)$$

$$d_E = \frac{v_D}{v_E + v_D} = \frac{v_O/k_L}{v_{IN} + \left(v_O/k_L\right)} = \frac{v_O}{v_{IN}k_L + v_O}$$

Transfer: Transformer current $i_{XI} = i_{LI} + k_L i_{LO} = k_L i_{XO}$ $\qquad i_{XO} = \frac{i_{LI}}{k_L} + i_{LO} = \frac{i_{XI}}{k_L}$

$$\therefore \quad P_O = v_O i_{O(AVG)} = v_O i_{XO(AVG)} d_D = v_O i_{XO(AVG)}(1 - d_E)$$

$$P_{IN} = v_{IN} i_{XI(AVG)} d_E = v_{IN} i_{XO(AVG)} k_L d_E = v_{IN}\left(\frac{i_{O(AVG)}}{d_D}\right) k_L d_E = P_O$$

$E_{IN} = E_X = E_O$ (↑ ↑ Ideal)

Variants:

High-Side Switches

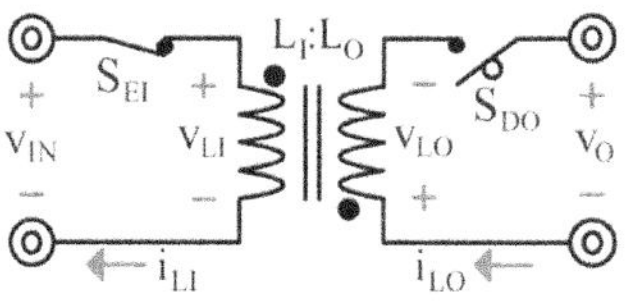

Low-Side Switches

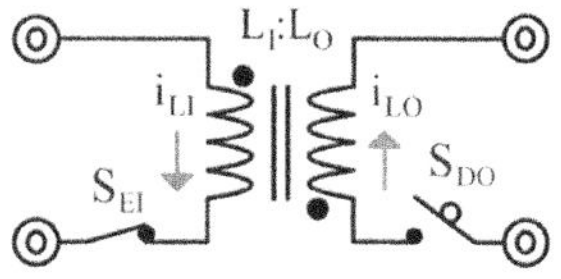

High-/Low-Side Switches

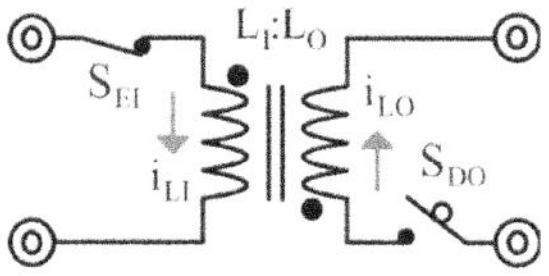

Low-/High-Side Switches

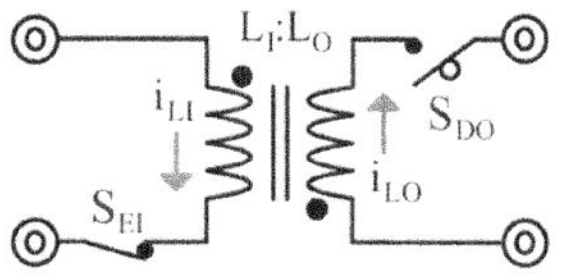

Challenge: Actual L_O cannot drain uncoupled L_I'

$\therefore$ L_I' & C_{SWI} exchange energy → v_{SWI} overshoots → v_{SWI} can break S_{EI}

Fix: Snubbers protect S_{EI}

Input Damped: R_{SI} burns E_{LC} $\therefore$ C_{SI} shorts below f_{LC}, but not below f_{SW}

$$Z_C = \frac{1}{sC_{SI}} \le R_{SI} \text{ past } f_{SI} = \frac{1}{2\pi R_{SI} C_{SI}}$$

$Z_C > R_{SI}$ below f_{SW} → $f_{SI} > f_{SW}$

$Z_C < R_{SI}$ below f_{LC} → $f_{SI} < f_{LC}$

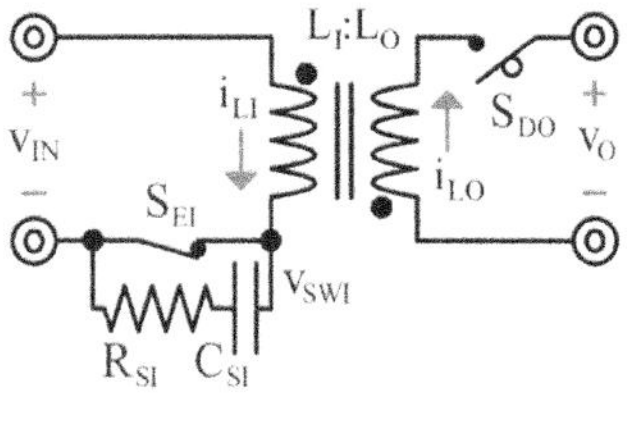

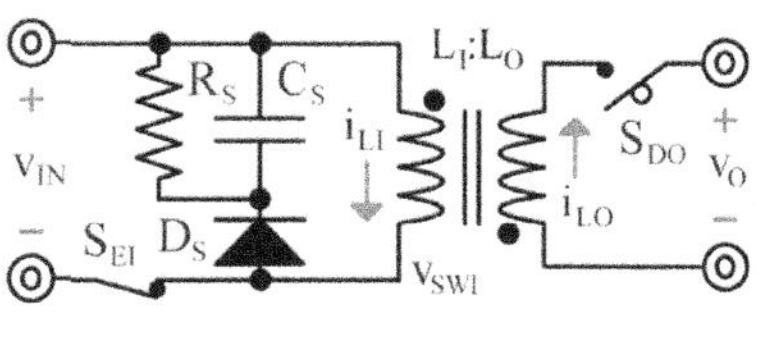

Input Clamped: D_S–C_S clamps v_{SWI} $\quad$ C_S accumulates q_C $\quad$ R_S leaks excess E_C

B. Asynchronous

Concept: Once energized, stored i_L can close drain diode automatically.

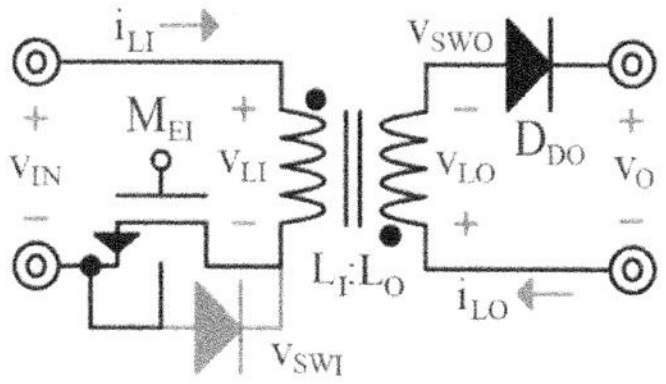

Drain Diode: Keep L_O grounded ∴ High-side D_{DO}

Energize FET: $K_N' > K_P$ ∴ Low-side NMOS M_{EI}

Block i_{IN} ∴ $v_{EI(B)}$ to Gnd

$$v_{LI(AVG)} \approx v_{IN}d_E' - \left(\frac{v_O + v_{DO}}{k_L}\right)d_D' = 0$$

$$v_{LO(AVG)} \approx v_{IN}k_L d_E' - (v_O + v_{DO})d_D' = 0$$

$$d_E' = \frac{v_D}{v_E + v_D} \approx \frac{(v_O + v_{DO})/k_L}{v_{IN} + (v_O + v_{DO})/k_L} = \frac{v_O + v_{DO}}{v_{IN}k_L + v_O + v_{DO}} > d_E$$

Conduction: D_{DO} blocks reverse $i_{L(D)}$ ∴ DCM

→ After t_C, $v_O + v_{DO}$ drains C_{SWO} into L_O

→ After t_C, $(v_O + v_{DO})/k_L$ drains C_{SWI} into L_I

} C_{SW}'s & L's exchange energy

C. Synchronous

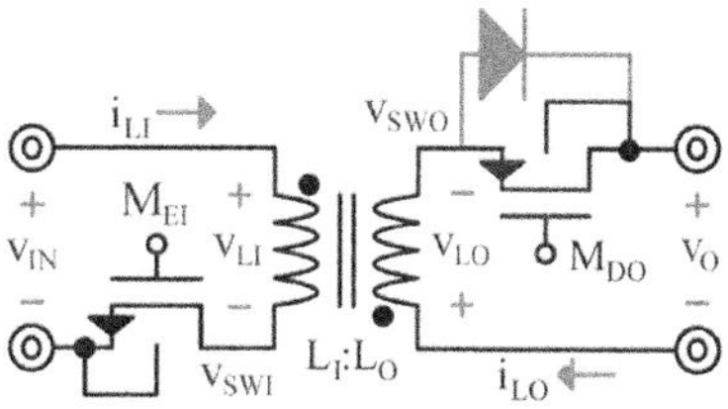

Energize FET: S_{EI} → M_{EI}

Drain FET: S_{DO} → High v_{SW} → PMOS

Block reverse i_O ∴ v_O to $v_{DO(B)}$

Caution: Keep FETs from closing simultaneously → Insert t_{DT} ∴ D_{DO} steers $i_{L(DT)}$

$$v_{LI(AVG)} \approx v_{IN}d_E'' - \left(\frac{v_O + v_{DO}}{k_L}\right)\left(\frac{2t_{DT}}{t_{SW}}\right) - \frac{v_O}{k_L}\left(\frac{t_D - 2t_{DT}}{t_{SW}}\right) = 0$$

$$v_{LO(AVG)} \approx v_{IN}k_L d_E'' - (v_O + v_{DO})\left(\frac{2t_{DT}}{t_{SW}}\right) - v_O\left(\frac{t_D - 2t_{DT}}{t_{SW}}\right) = 0$$

$$d_E'' = \frac{v_D}{v_E + v_D} \approx \frac{v_O/k_L}{v_{IN} + (v_O/k_L)} + \left[\frac{v_{DO}/k_L}{v_{IN} + (v_O/k_L)}\right]\left(\frac{2t_{DT}}{t_{SW}}\right)$$

$$= \frac{v_O}{v_{IN}k_L + v_O} + \left(\frac{v_{DO}}{v_{IN}k_L + v_O}\right)\left(\frac{2t_{DT}}{t_{SW}}\right) = d_E + \left(\frac{v_{DO}}{v_{IN}k_L + v_O}\right)\left(\frac{2t_{DT}}{t_{SW}}\right) > d_E$$

Chapter 4. Power Losses

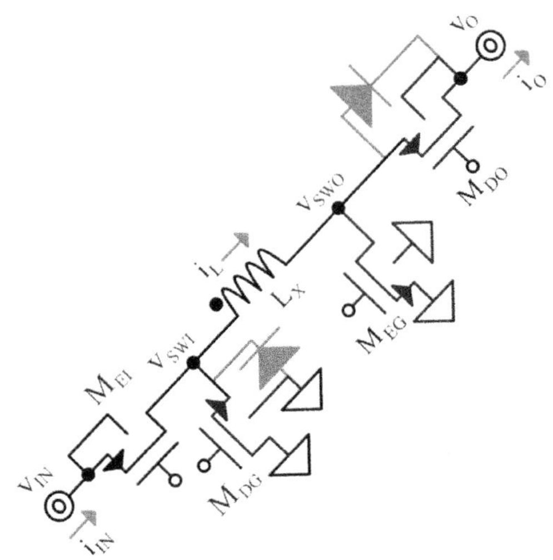

4.1. Power Conversion

$d_{IN} \equiv$ Input Duty Cycle $d_O \equiv$ Output Duty Cycle

Fractional Losses $\sigma_{LOSS} \equiv$ Fraction of P_{IN} that P_{LOSS} consumes

$$\sigma_{LOSS} = \frac{P_{LOSS}}{P_{IN}} = \frac{P_{LOSS}}{i_{IN(AVG)} v_{IN}} = \frac{P_{LOSS}}{i_{L(AVG)} d_{IN} v_{IN}} = \frac{P_{LOSS}}{\left(i_{O(AVG)}/d_O\right) d_{IN} v_{IN}}$$

Power-Conversion Efficiency $\eta_C \equiv$ Fraction of P_{IN} that P_O delivers

$$\eta_C \equiv \frac{P_O}{P_{IN}} = \frac{P_O}{P_O + P_{LOSS}} = \frac{P_{IN} - P_{LOSS}}{P_{IN}} = 1 - \frac{P_{LOSS}}{P_{IN}} = 1 - \sigma_{LOSS}$$

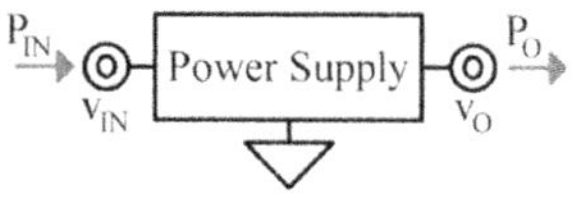

Voltage Regulators: Regulate v_O Supply $i_{LD} = i_{O(AVG)}$

Battery Chargers: Monitor v_B Regulate & Supply $i_{O(AVG)}$

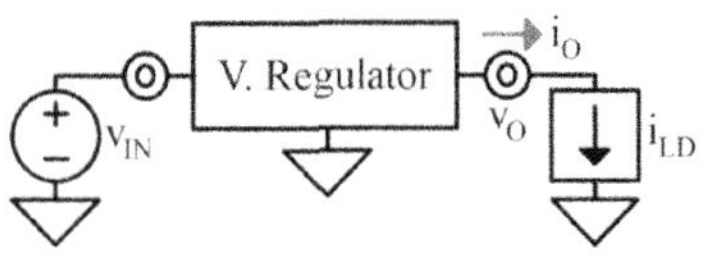

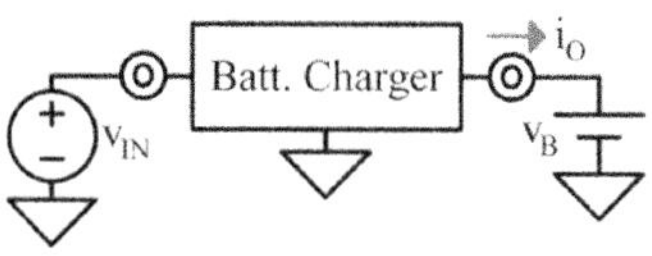

$$\eta_C = \frac{i_{O(AVG)} v_O}{i_{O(AVG)} v_O + P_{LOSS}}$$

Source: Good → $P_{IN(MAX)} >> P_{LD(MAX)} + P_{LOSS}$ → Low R_S

∴ High η_C (low σ_{LOSS}) saves energy

Poor → $P_{IN(MAX)} < P_{LD(MAX)} + P_{LOSS}$ → R_S limits $P_{IN(MAX)}$

∴ Maxing P_O is more important than reducing σ_{LOSS}

→ $P_{O(MPP)} \equiv$ Maximum-Power Point (MPP)

$P_{O(MPP)} \equiv P_{O(MAX)}$ when η_C peaks at $P_{O(MPP)}$

Energy Harvesters: Ambient source is poor

∴ MPP Trackers (MPPT) track MPP

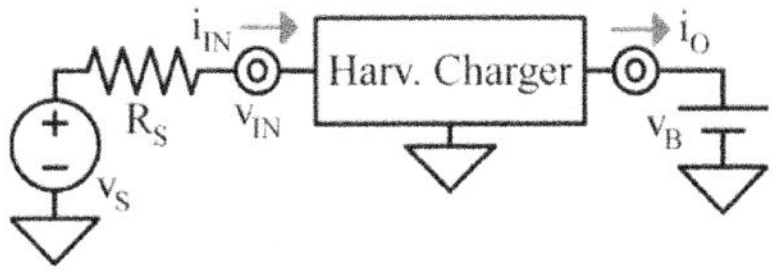

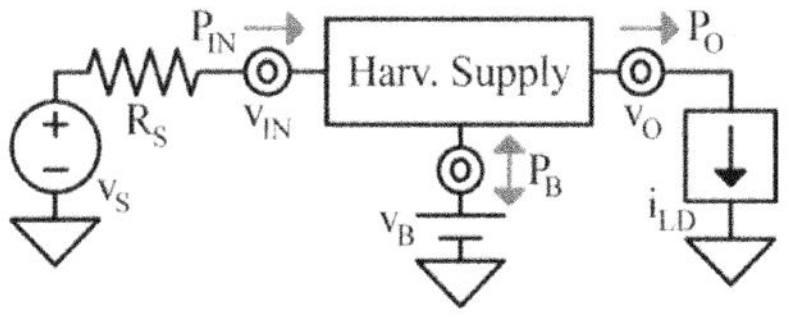

4.2. Operating Mechanics

Continuous Conduction

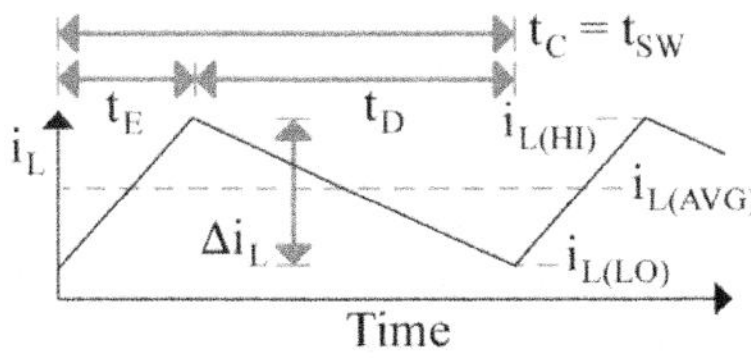

Discontinuous Conduction

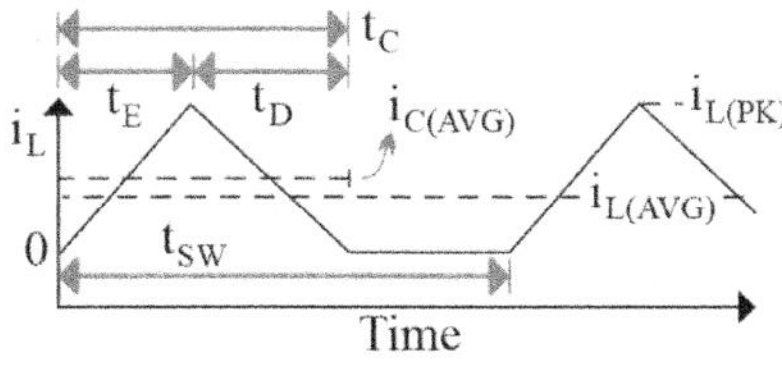

Discontinuous i_L:

$$i_{L(AVG)} = i_{C(AVG)}\left(\frac{t_C}{t_{SW}}\right) = \left(\frac{i_{L(PK)}}{2}\right)\left(\frac{t_C}{t_{SW}}\right) = \frac{i_{L(PK)}{}^2 L_X}{2 d_E v_E t_{SW}}$$

$$d_E = \frac{v_D}{v_E + v_D} \qquad t_C = \frac{t_E}{d_E} = \left(\frac{i_{L(PK)}}{d_E}\right)\left(\frac{L_X}{v_E}\right)$$

$$i_{L(PK)} = \sqrt{2 d_E \left(\frac{v_E}{L_X}\right) t_{SW} i_{L(AVG)}} = \sqrt{2\left(\frac{v_E \,||\, v_D}{L_X}\right) t_{SW}\left(\frac{i_{O(AVG)}}{d_O}\right)}$$

4.3. Magnetic (Core) Loss

i_L magnetizes/drains core

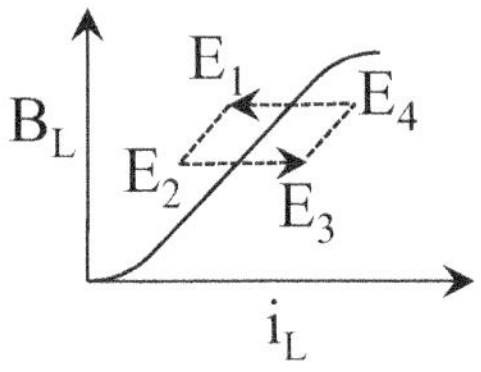

Solenoid:

# of Turns

$$B_L = \mu_C\left(\frac{N_L}{l_L}\right)i_L = \left(\frac{L_X}{N_L A_L}\right)i_L$$

Permeability Length Cross-Sectional Area

Molecular interactions resist magnetic re-orientation → Lose heat energy

$\therefore \quad E_{IN} = E_L + E_{LOSS(IN)} = \Delta E_{34} \qquad E_O = E_L - E_{LOSS(O)} = \Delta E_{12} \qquad \sigma_C \approx 3\%\text{–}5\%$

$$\Delta E_{AB} = \left(\frac{1}{2}\right)L_X\left[\left(i_{LAB(AVG)} + \frac{\Delta i_L}{2}\right)^2 - \left(i_{LAB(AVG)} - \frac{\Delta i_L}{2}\right)^2\right] = \left(\frac{1}{2}\right)L_X\left(2i_{LAB(AVG)}\Delta i_L\right)$$

→ $E_C = E_{IN} - E_O = \Delta E_{34} - \Delta E_{12} = L_X\Delta i_{L(AVG)}\Delta i_L \approx L_X(k_C\Delta i_L)\Delta i_L \equiv R_M\Delta i_L^2$

$P_C = E_{LOSS}f_{SW} \approx k_C L_X \Delta i_L^2 f_{SW}$ Examples: $k_{C(AIR)} \approx 0$, $k_{C(X)} \approx 3\%$

4.4. Ohmic Loss

Average Power: $P_{AX} \equiv P_{A(AVG)}\Big|_0^{t_X} = \frac{1}{t_X}\int_0^{t_X} P_A dt = \frac{1}{t_X}\int_0^{t_X} i_A v_A dt$

A. Ohmic Power

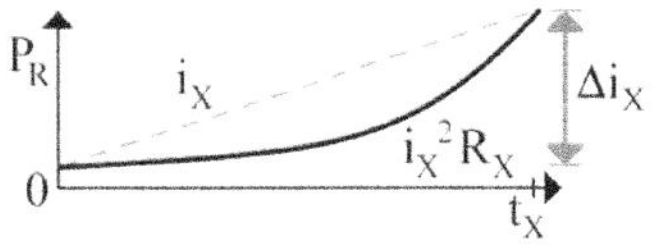

$$P_{RX} = \frac{1}{t_X}\int_0^{t_X} i_X v_R dt = \left(\frac{1}{t_X}\int_0^{t_X} i_X^2 dt\right)R_X = i_{X(RMS)}^2 R_X$$

Triangular Current:

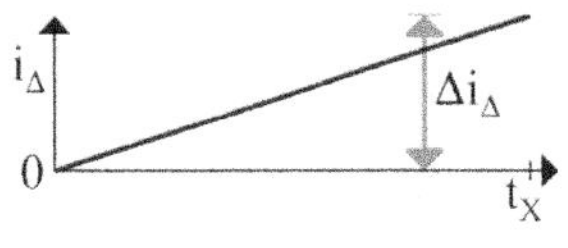

$$i_{\Delta(RMS)} = \sqrt{\frac{1}{t_X}\int_0^{t_X}\left(\frac{\Delta i_\Delta t}{t_X}\right)^2 dt} = \sqrt{\left(\frac{\Delta i_\Delta^2}{t_X^3}\right)\left(\frac{t_X^3}{3}\right)} = \frac{\Delta i_\Delta}{\sqrt{3}}$$

Alternating Current:

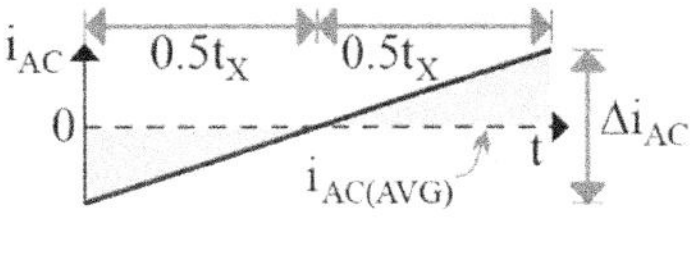

$$i_{AC(RMS)} \equiv i_{AC(RMS)}\Big|_{t_X} = i_{AC(RMS)}\Big|_{0.5t_X} = i_{\Delta(RMS)}\Big|_{0.5t_X} = \frac{0.5\Delta i_{AC}}{\sqrt{3}}$$

Power Theorem:

$$i_X = i_{X(MIN)} + \left(\frac{\Delta i_X}{t_X}\right)t$$

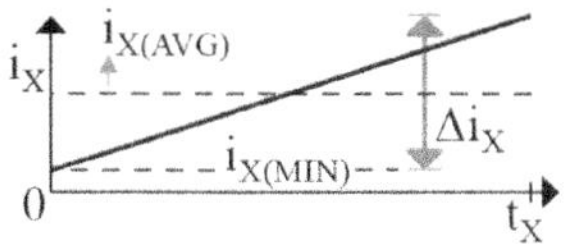

$$i_{X(MIN)} = i_{X(AVG)} - \frac{\Delta i_X}{2}$$

$$i_{X(RMS)}{}^2 = \frac{1}{t_X}\int_0^{t_X} i_X{}^2 dt$$

$$= \frac{1}{t_X}\left[i_{X(MIN)}{}^2 t + \left(\frac{\Delta i_X{}^2}{t_X{}^2}\right)\left(\frac{t^3}{3}\right) + 2i_{X(MIN)}\left(\frac{\Delta i_X}{t_X}\right)\left(\frac{t^2}{2}\right)\right]\Bigg|_0^{t_X}$$

$$= \frac{1}{t_X}\left[\left(i_{X(AVG)} - \frac{\Delta i_X}{2}\right)^2 t_X + \Delta i_X{}^2\left(\frac{t_X}{3}\right) + \left(i_{X(AVG)} - \frac{\Delta i_X}{2}\right)\Delta i_X t_X\right]$$

$$= i_{X(AVG)}{}^2 + \left(\frac{\Delta i_X}{2}\right)^2\left(1 + \frac{4}{3} - 2\right) = i_{X(AVG)}{}^2 + \left(\frac{0.5\Delta i_X}{\sqrt{3}}\right)^2 = i_{X(AVG)}{}^2 + i_{AC(RMS)}{}^2$$

$$\therefore \quad P_R = P_{RX}\left(\frac{t_X}{t_{SW}}\right) = i_{X(RMS)}{}^2 R_X\left(\frac{t_X}{t_{SW}}\right) = \left(i_{X(AVG)}{}^2 + i_{AC(RMS)}{}^2\right)R_X\left(\frac{t_X}{t_{SW}}\right)$$

B. Continuous Conduction

Switched L_X: $P_{RL} \approx i_{L(AVG)}{}^2 R_{L(DC)} + i_{AC(RMS)}{}^2 R_{L(AC)} = \left(\frac{i_{O(AVG)}}{d_O}\right)^2 R_{L(DC)} + \left(\frac{0.5\Delta i_L}{\sqrt{3}}\right)^2 R_{L(AC)}$

Energize Resistances:

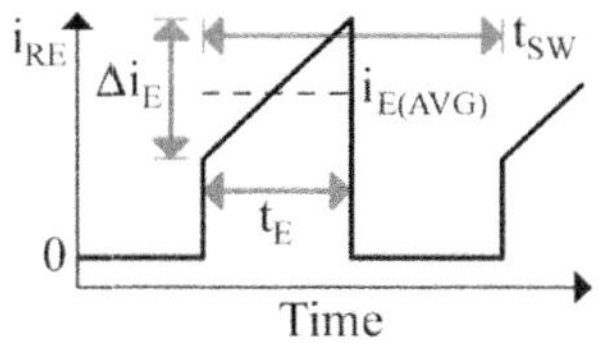

$$P_{RE} = i_{RE(RMS)}{}^2 R_E = \left(i_{E(AVG)}{}^2 + i_{AC(RMS)}{}^2\right)R_E\left(\frac{t_E}{t_{SW}}\right)$$

$$= \left[\left(\frac{i_{O(AVG)}}{d_O}\right)^2 + \left(\frac{0.5\Delta i_L}{\sqrt{3}}\right)^2\right]\left(R_{EI} + R_{EG}\right)d_E$$

Drain Resistances:

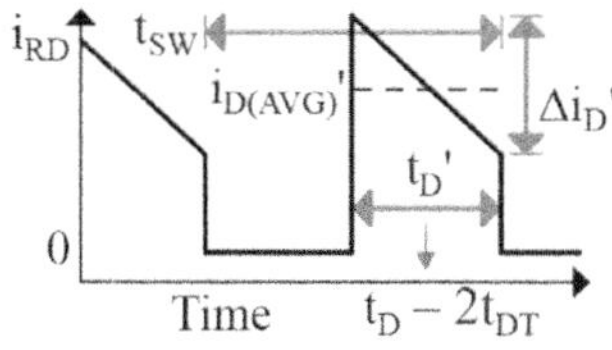

$$P_{RD} = i_{RD(RMS)}{}^2 R_D = \left(i_{D(AVG)}{}^2 + i_{AC(RMS)}{}^2\right)R_D\left(\frac{t_D - 2t_{DT}}{t_{SW}}\right)$$

$$\approx \left[\left(\frac{i_{O(AVG)}}{d_O}\right)^2 + \left(\frac{0.5\Delta i_L}{\sqrt{3}}\right)^2\right]\left(R_{DG} + R_{DO}\right)\left(d_D - \frac{2t_{DT}}{t_{SW}}\right)$$

Example: P_{RE} & σ_{RE} for M_{EG} in an ideal buck–boost in CCM when $v_{IN} = 2$ V, $v_O = 4$ V, $i_{O(AVG)} = 250$ mA, $L_X = 10$ µH, $t_{SW} = 1$ µs, $R_{EG} = 200$ mΩ.

Solution:

$$d_{IN} = d_E = \frac{v_O}{v_{IN} + v_O} = \frac{4}{2+4} = 67\% \quad \therefore \quad d_O = d_D = 1 - d_E = 33\%$$

$$\Delta i_L = \left(\frac{v_E}{L_X}\right) d_E t_{SW} = \left(\frac{2}{10\mu}\right)(67\%)(1\mu) = 130 \text{ mA}$$

$$P_{RE} = \left[\left(\frac{i_{O(AVG)}}{d_O}\right)^2 + \left(\frac{0.5\Delta i_L}{\sqrt{3}}\right)^2\right] R_{EG} d_E$$

Note: Ideal d_E excludes v_R

→ v_R raises d_E → P_{RE}, P_{IN}

$$= \left\{\left(\frac{250m}{33\%}\right)^2 + \left[\frac{0.5(130m)}{\sqrt{3}}\right]^2\right\}(200m)(67\%) = 77 \text{ mW}$$

$$\sigma_{RE} = \frac{P_{RE}}{\left(i_{O(AVG)}/d_O\right) d_{IN} v_{IN}} = \frac{77m}{(250m/33\%)(67\%)(2)} = 7.6\%$$

SPICE Simulation

```
* Ideal Buck-Boost with REG in CCM
vde de 0 dc=0 pulse 0 1 0 1n 1n 684n 1u
vin vin 0 dc=2
sei vin vswi de 0 sw1v
ddg 0 vswi idiode
lx vswi vswo 10u
seg vswo 0 de 0 sw1v200m
ddo vswo vo idiode
vo vo 0 dc=4
.ic i(lx)=700m
.model sw1v vswitch roff=1e12 ron=1m voff=450m von=550m
.model sw1v200m vswitch roff=1e12 ron=200m voff=450m von=550m
.model idiode d is=1f n=0.001
```

```
.tran 1u
.end
* Plot –iEG
* Extract iEG(RMS)
* Square iEG(RMS)
* Multiply times REG
```

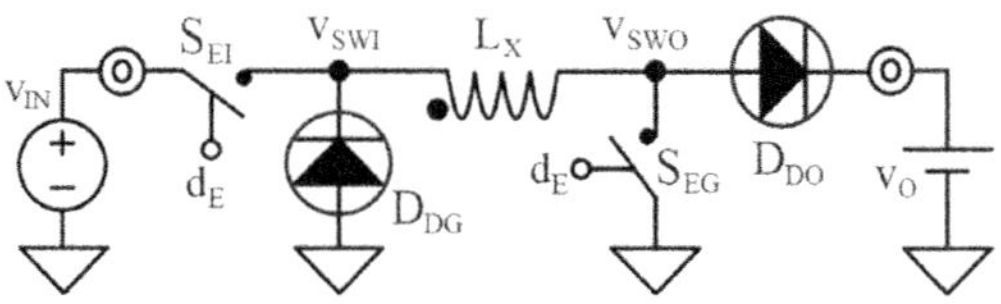

Output Capacitor:

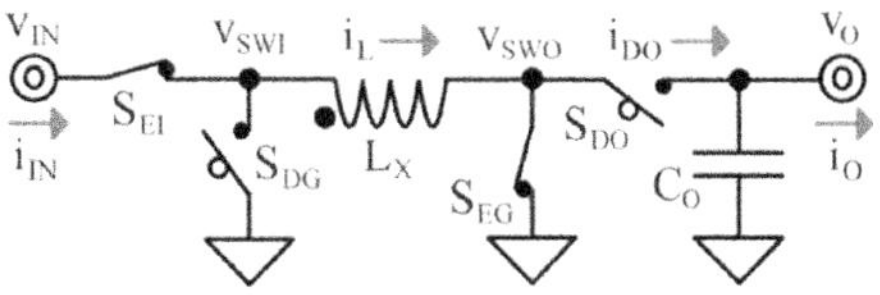

C_O supplies i_{LD} when S_{DO} opens.

Duty-Cycled v_O: $d_O = d_D$ $\quad P_{RC(DO)} = P_{RC}\big|_{t_E} + P_{RC}\big|_{t_{DO}}$ $\quad$ Static $i_O \equiv i_{O(AVG)}$

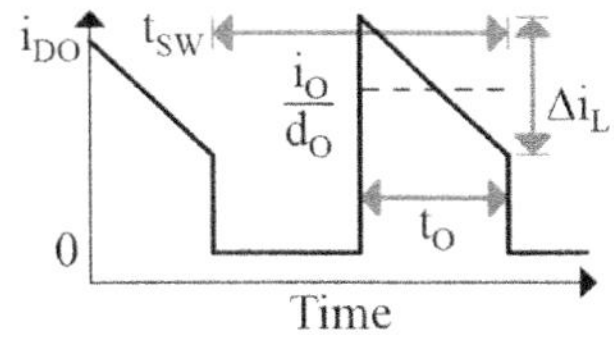

$$= i_O{}^2 R_C\left(\frac{t_E}{t_{SW}}\right) + \left[\left(i_{D(AVG)} - i_O\right)^2 + i_{AC(RMS)}{}^2\right] R_C\left(\frac{t_O}{t_{SW}}\right)$$

$$= i_O{}^2 R_C d_E + \left[\left(\frac{i_O}{d_O} - i_O\right)^2 + \left(\frac{0.5\Delta i_L}{\sqrt{3}}\right)^2\right] R_C d_D$$

Buck: $d_O = 1$

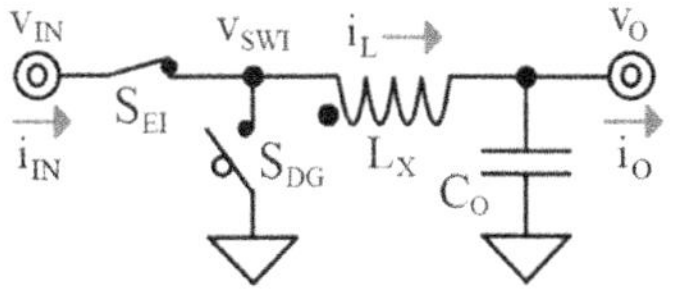

$$P_{RC(BK)} = i_{C(RMS)}{}^2 R_C = (i_{C(AVG)}{}^2 + i_{AC(RMS)}{}^2) R_C$$

$$= \left[0^2 + \left(\frac{0.5\Delta i_L}{\sqrt{3}}\right)^2\right] R_C << P_{RC(DO)}$$

Example: P_{RC} & σ_{RC} for C_O in an ideal boost in CCM when $v_{IN} = 2$ V, $v_O = 4$ V, $i_O = 250$ mA, $L_X = 10$ μH, $t_{SW} = 1$ μs, $R_C = 200$ mΩ.

Solution: Boost $\therefore$ $d_{IN} = 1$

$$d_E = \frac{v_O - v_{IN}}{v_O} = \frac{4-2}{4} = 50\% \quad \therefore \quad d_O = d_D = 1 - d_E = 50\%$$

$$\Delta i_L = \left(\frac{v_{IN}}{L_X}\right) d_E t_{SW} = \left(\frac{2}{10\mu}\right)(50\%)(1\mu) = 100 \text{ mA}$$

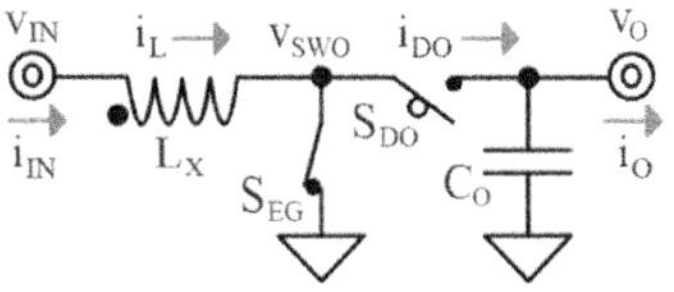

$$P_{RC} = i_O{}^2 R_C d_E + \left[\left(\frac{i_O}{d_O} - i_O\right)^2 + \left(\frac{0.5\Delta i_L}{\sqrt{3}}\right)^2\right] R_C d_O = 13 \text{ mW}$$

$$= (250m)^2(200m)(50\%) + \left\{\left(\frac{250m}{50\%} - 250m\right)^2 + \left[\frac{0.5(100m)}{\sqrt{3}}\right]^2\right\}(200m)(50\%)$$

$$\sigma_{RC} = \frac{P_{RC}}{(i_O/d_O)d_{IN}v_{IN}} = \frac{13m}{(250m/50\%)(1)(2)} = 1.3\%$$

C. Discontinuous Conduction

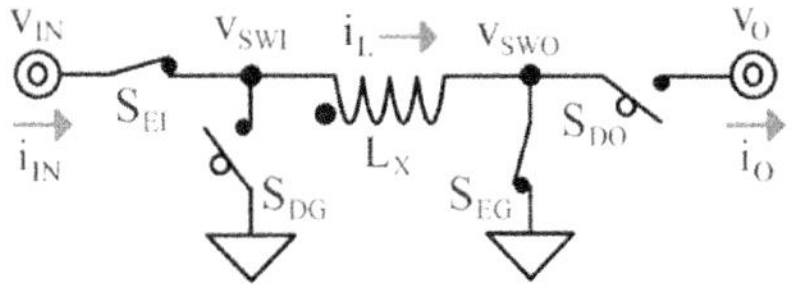

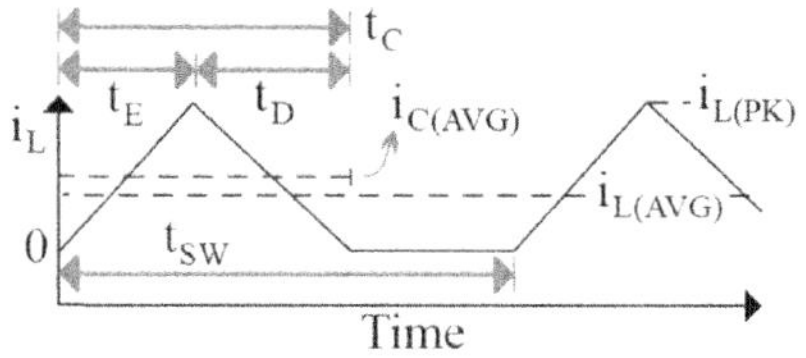

Switched Inductor:

$$P_{RL} = i_{L(RMS)}{}^2 R_{L(AC)} = i_{C(RMS)}{}^2 R_{L(AC)} \left(\frac{t_C}{t_{SW}}\right) = \left(\frac{i_{L(PK)}}{\sqrt{3}}\right)^2 R_{L(AC)} \left(\frac{t_C}{t_{SW}}\right)$$

Energize Resistances:

$$P_{RE} = i_{RE(RMS)}{}^2 R_E = i_{E(RMS)}{}^2 R_E \left(\frac{t_E}{t_{SW}}\right) = \left(\frac{i_{L(PK)}}{\sqrt{3}}\right)^2 (R_{EI} + R_{EG}) d_E \left(\frac{t_C}{t_{SW}}\right)$$

Drain Resistances:

$$P_{RD} = i_{RD(RMS)}{}^2 R_D = i_{D(RMS)}{}^2 R_D \left(\frac{t_D}{t_{SW}}\right) = \left(\frac{i_{L(PK)}}{\sqrt{3}}\right)^2 (R_{DG} + R_{DO}) d_D \left(\frac{t_C}{t_{SW}}\right)$$

Output Capacitor:

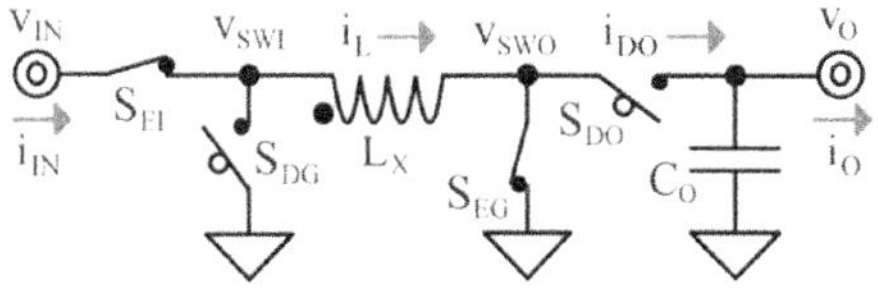

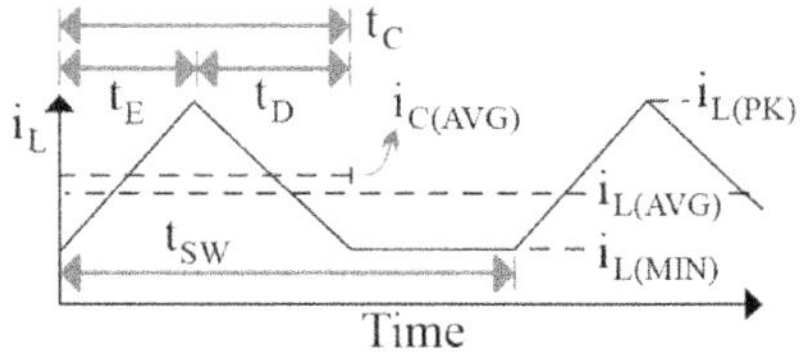

Duty-Cycled v_O: $d_O = d_D$

$$P_{RC(DO)} = P_{RC}\big|_{t_{SW}-t_O} + P_{RC}\big|_{t_O} = i_O{}^2 R_C \left(\frac{t_{SW} - t_O}{t_{SW}}\right) + \left[\left(i_{D(AVG)} - i_O\right)^2 + i_{AC(RMS)}{}^2\right] R_C \left(\frac{t_O}{t_{SW}}\right)$$

$$= i_O{}^2 R_C \left[1 - d_D\left(\frac{t_C}{t_{SW}}\right)\right] + \left[\left(\frac{i_{L(PK)}}{2} - i_O\right)^2 + \left(\frac{0.5 i_{L(PK)}}{\sqrt{3}}\right)^2\right] R_C d_D \left(\frac{t_C}{t_{SW}}\right)$$

Buck: $d_O = 1$

$$P_{RC(BK)} = P_{RC}\big|_{t_{SW}-t_C} + P_{RC}\big|_{t_C} = i_O{}^2 R_C \left(\frac{t_{SW} - t_C}{t_{SW}}\right) + \left[\left(i_{C(AVG)} - i_O\right)^2 + i_{AC(RMS)}{}^2\right] R_C \left(\frac{t_C}{t_{SW}}\right)$$

$$= i_O{}^2 R_C \left(1 - \frac{t_C}{t_{SW}}\right) + \left[\left(\frac{i_{L(PK)}}{2} - i_O\right)^2 + \left(\frac{0.5 i_{L(PK)}}{\sqrt{3}}\right)^2\right] R_C \left(\frac{t_C}{t_{SW}}\right)$$

Example: P_{RC} & σ_{RC} for C_O in the ideal boost from previous example in DCM when $i_O = 10$ mA.

Solution: $d_{IN} = 1$, $d_E = 50\%$, $d_O = d_D = 50\%$ from previous example

$$i_{L(PK)} = \sqrt{2d_E t_{SW}\left(\frac{v_E}{L_X}\right)\left(\frac{i_O}{d_O}\right)} = \sqrt{2(50\%)(1\mu)\left(\frac{2}{10\mu}\right)\left(\frac{10m}{50\%}\right)} = 63\ \text{mA}$$

$$t_C = \left(\frac{i_{L(PK)}}{d_E}\right)\left(\frac{L_X}{v_E}\right) = \left(\frac{63m}{50\%}\right)\left(\frac{10\mu}{2}\right) = 630\ \text{ns}$$

$$\sigma_{RC} = \frac{P_{RC}}{(i_O/d_O)d_{IN}v_{IN}} = 0.2\%$$

$$P_{RC} = i_O^2 R_C\left[1 - d_D\left(\frac{t_C}{t_{SW}}\right)\right] + \left[\left(\frac{i_{L(PK)}}{2} - i_O\right)^2 + \left(\frac{i_{L(PK)}}{2\sqrt{3}}\right)^2\right] R_C d_D\left(\frac{t_C}{t_{SW}}\right)$$

$$= (10m)^2(200m)\left[1 - 50\%\left(\frac{630n}{1\mu}\right)\right]$$

$$+ \left[\left(\frac{63m}{2} - 10m\right)^2 + \left(\frac{63m}{2\sqrt{3}}\right)^2\right](200m)(50\%)\left(\frac{630n}{1\mu}\right) = 64\ \mu\text{W}$$

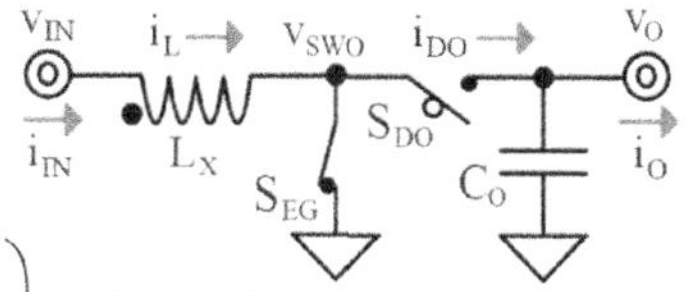

4.5. Diode Loss

Conduction Power: $P_{VX} = \frac{1}{t_X}\int_0^{t_X} i_X v_X dt \approx \left(\frac{1}{t_X}\int_0^{t_X} i_X dt\right) v_X = i_{X(AVG)} v_X$

$$i_{X(AVG)} = \frac{1}{t_X}\int_0^{t_X} i_X dt = \frac{1}{t_X}\int_0^{t_X}\left[i_{X(MIN)} + \left(\frac{\Delta i_X}{t_X}\right)t\right]dt = i_{X(MIN)} + \frac{\Delta i_X}{2}$$

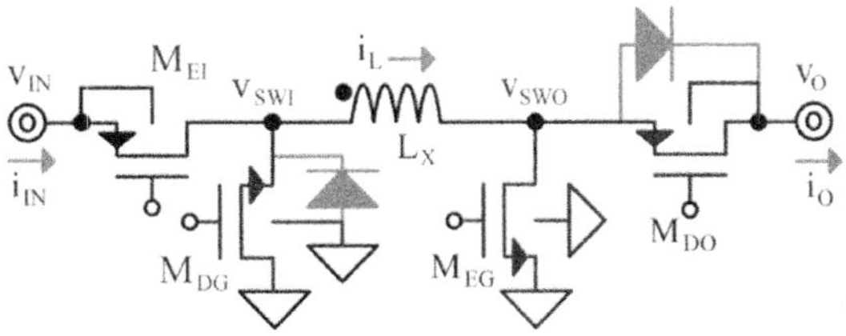

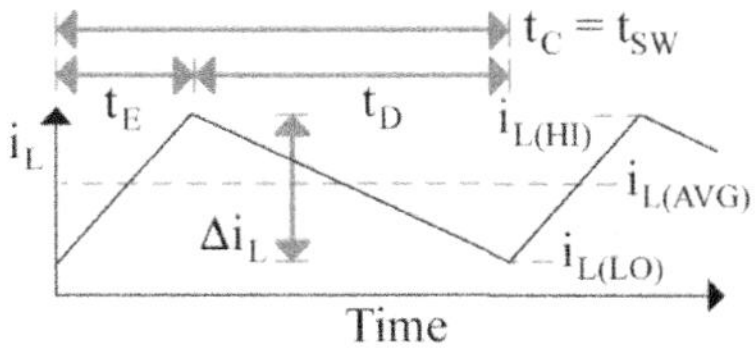

Continuous Conduction: $P_{DT} \approx i_{L(HI)}(v_{DG} + v_{DO})\left(\frac{t_{DT}}{t_{SW}}\right) + i_{L(LO)}(v_{DG} + v_{DO})\left(\frac{t_{DT}}{t_{SW}}\right)$

$$\approx 2i_{L(AVG)}(v_{DG} + v_{DO})\left(\frac{t_{DT}}{t_{SW}}\right) = \left(\frac{i_{O(AVG)}}{d_O}\right)(v_{DG} + v_{DO})\left(\frac{2t_{DT}}{t_{SW}}\right)$$

Discontinuous Conduction: $P_{DT} \approx i_{L(PK)}(v_{DG} + v_{DO})\left(\frac{t_{DT}}{t_{SW}}\right)$

Example: P_{DT} & σ_{DT} in an ideal sync. buck in DCM when $v_{IN} = 4$ V, $v_O = 2$ V,

$i_{O(AVG)} = 10$ mA, $L_X = 10$ µH, $t_{SW} = 1$ µs, $t_{DT} = 50$ ns, D_{DG} drops 700 mV.

Solution: Buck $\therefore$ $d_O = 1$

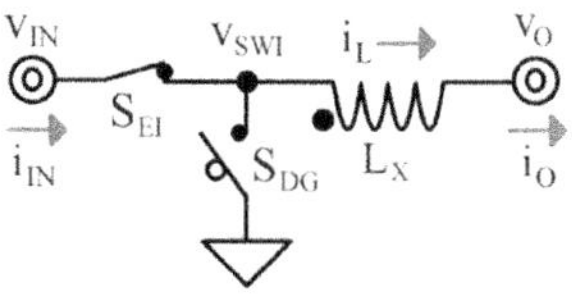

$$d_{IN} = d_E = \frac{v_O}{v_{IN}} + \left(\frac{v_{DG}}{v_{IN}}\right)\left(\frac{t_{DT}}{t_C}\right) \approx \frac{v_O}{v_{IN}} = \frac{2}{4} = 50\%$$

$$i_{L(PK)} = \sqrt{2d_E t_{SW}\left(\frac{v_E}{L_X}\right) i_{O(AVG)}} \approx \sqrt{2(50\%)(1\mu)\left(\frac{4-2}{10\mu}\right)(10m)} = 45 \text{ mA}$$

$$P_{DT} \approx i_{L(PK)} v_{DG}\left(\frac{t_{DT}}{t_{SW}}\right) \approx (45m)(700m)\left(\frac{50n}{1\mu}\right) = 1.6 \text{ mW}$$

$$\sigma_{DT} = \frac{P_{DT}}{\left(i_{O(AVG)}/d_O\right) d_{IN} v_{IN}} \approx \frac{1.6m}{(10m/1)(50\%)(4)} = 8.0\%$$

$\approx \sqrt{1.9\%}$ Higher

$$t_C = \left(\frac{i_{L(PK)}}{d_E}\right)\left(\frac{L_X}{v_E}\right) \approx 450 \text{ ns} \quad \rightarrow \quad \Delta d_E = \left(\frac{v_{DG}}{v_{IN}}\right)\left(\frac{t_{DT}}{t_C}\right) \approx 1.9\%$$

4.6. i_{DS}–v_{DS} Overlap Loss: A. Closing Switch

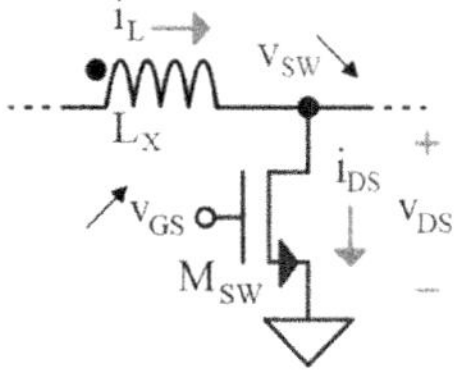

Transition: i_{DS} climbs quadratically when $v_{GS} > V_{T0}$.

v_{DS} falls when $i_{DS} > i_L + i_{GD} + i_{SW} \approx i_L$.

i_P slews C_{GD} at v_{GS} needed to sustain $i_{DS} \approx i_L$.

M_{SW} transitions to triode when $v_{DS} < v_{GS}$.

$\therefore$ C_{GD} adds $0.5C_{CH}$ → Slows v_{SW}'s transition

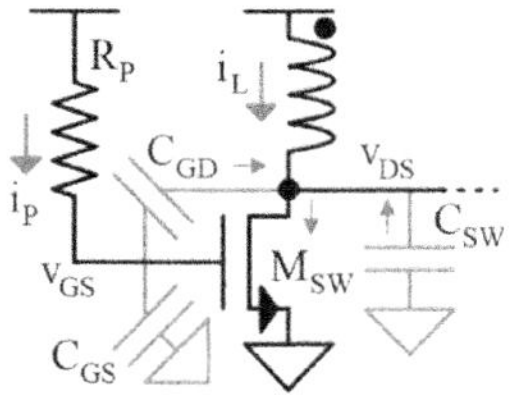

Power: $$P_I = \frac{1}{t_I}\int_0^{t_I} i_{DS} v_{DS} dt = \left(\frac{1}{t_I}\int_0^{t_I} i_{DS} dt\right)\Delta v_{SW} \approx \left[\frac{1}{t_I}\int_0^{t_I}\left(\frac{i_L}{t_I^2}\right) t^2 dt\right]\Delta v_{SW} \approx \left(\frac{i_L}{3}\right)\Delta v_{SW}$$

$$P_V = \frac{1}{t_V}\int_0^{t_V} i_{DS} v_{DS} dt \approx i_L\left(\frac{1}{t_V}\int_0^{t_V} v_{DS} dt\right) \approx i_L\left(\frac{\Delta v_{SW}}{2}\right)$$

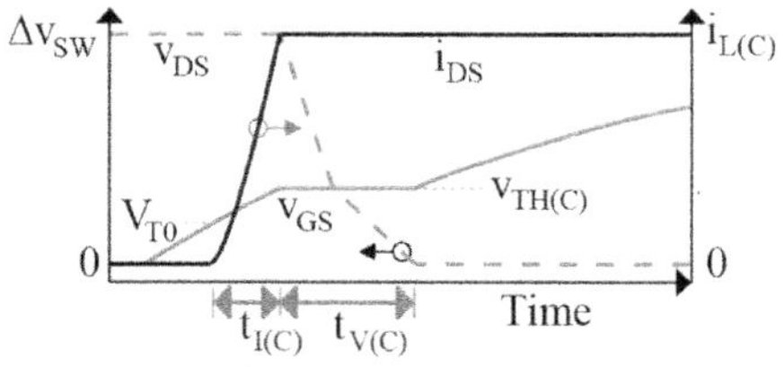

$$P_{IV} = P_I\left(\frac{t_I}{t_{SW}}\right) + P_V\left(\frac{t_V}{t_{SW}}\right) \approx i_L \Delta v_{SW}\left(\frac{t_I}{3t_{SW}} + \frac{t_V}{2t_{SW}}\right)$$

→ i_{DS} transitions before v_{DS}.

Delays:

$t_{I(C)}$: R_P charges C_G to v_{GS} needed for i_L: $v_{TH(C)} \approx V_{T0} + v_{DS(SAT)}\Big|_{i_{L(C)}} \approx V_{T0} + \sqrt{\dfrac{2i_{L(C)}}{K'(W/L)}}$

RC time to charge Δv_X: $t_X = \tau_{RC} \ln\left(\dfrac{v_{DD}}{v_{DD} - \Delta v_X}\right)$

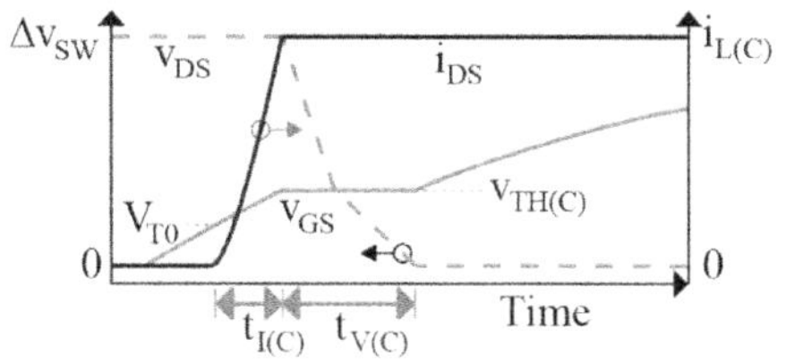

Saturated Inversion

$$\tau_{RC(C)} = R_P\left(C_{GS} + C_{GD}\right) \approx R_P\left[2C_{OL} + \left(\frac{2}{3}\right)C_{CH}\right]$$

i_{DS} rises when v_{GS} rises from V_{T0} to $v_{TH(C)}$: $t_{I(C)} \approx t_{TH(C)} - t_{T0} = \tau_{RC(C)} \ln\left(\dfrac{v_{DD} - V_{T0}}{v_{DD} - v_{TH(C)}}\right)$

$t_{V(C)}$: i_P slews C_{GD} at v_{GS} needed for i_L Sat.-to-tri. when $v_{DS} \leq v_{GS}$ → Avg $0.5C_{CH}$

$$t_{V(C)} \approx \left(\frac{\Delta v_C}{i_P}\right)C_{GD} \approx \left(\frac{R_P}{v_{DD} - v_{GS}}\right)\left[C_{OL}\Delta v_{SW} + \left(\frac{0.5C_{CH}}{2}\right)v_{GS}\right]$$

$$\approx \left(\frac{R_P}{v_{DD} - v_{TH(C)}}\right)\left[C_{OL}\Delta v_{SW} + \left(\frac{C_{CH}}{4}\right)v_{TH(C)}\right]$$

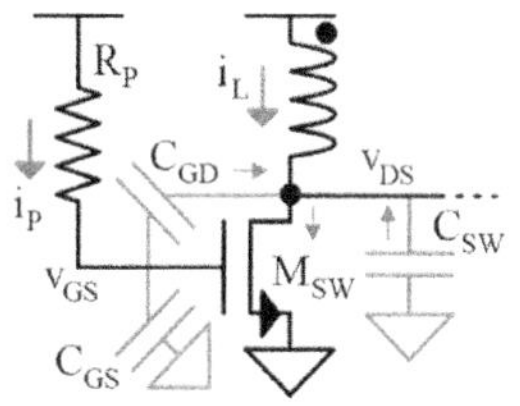

Example: P_{IV} & σ_{IV} for M_{EG} in an ideal synchronous boost in CCM when M_{EG} closes, $v_{IN} = 2$ V, $v_{DD} = v_O = 4$ V, $i_{O(AVG)} = 250$ mA, $L_X = 10$ µH, $t_{SW} = 1$ µs, D_{DO} drops 800 mV, $t_{DT} = 50$ ns, $W_{EG} = 50$ mm, $L_{EG} = 250$ nm, $L_{OL} = 30$ nm, $V_{TN0} = 400$ mV, $K_N' = 200$ µA/V², $C_{OX}'' = 6.9$ fF/µm², $R_P = 100$ Ω.

Solution: $L_{CH} = L_{EG} - 2L_{OL} = 250n - 2(30n) = 190$ nm

$$d_E \approx \frac{v_O - v_{IN}}{v_O} + \left(\frac{v_{DO}}{v_O}\right)\left(\frac{2t_{DT}}{t_{SW}}\right) = 52\% \quad \therefore \quad d_O = d_D = 1 - d_E \approx 48\%$$

$$\Delta i_L = \left(\frac{v_E}{L_X}\right)d_E t_{SW} \approx \left(\frac{2}{10\mu}\right)(52\%)(1\mu) = 100 \text{ mA}$$

M_{EG} closes when $i_L = i_{L(LO)} = \dfrac{i_{O(AVG)}}{d_O} - \dfrac{\Delta i_L}{2} \approx \dfrac{250m}{48\%} - \dfrac{100m}{2} = 470$ mA

$$v_{TH(C)} = V_{TN0} + v_{DS(SAT)} \approx V_{TN0} + \sqrt{\frac{2i_{L(LO)}}{K_N'(W_{CH}/L_{CH})}} = 400m + 130m = 530 \text{ mV}$$

$$C_{OL} = C_{OX}''W_{CH}L_{OL} = (6.9m)(50m)(30n) = 10\ pF$$

$$C_{CH} = C_{OX}''W_{CH}L_{CH} = (6.9m)(50m)(190n) = 66\ pF$$

$$\tau_{RC(C)} = R_P\left[2C_{OL} + \left(\frac{2}{3}\right)C_{CH}\right] = (100)\left[2(10p) + \left(\frac{2}{3}\right)(66p)\right] = 6.4\ ns$$

$$t_{I(C)} \approx \tau_{RC(C)} \ln\left(\frac{v_{DD} - V_{T0}}{v_{DD} - v_{TH(C)}}\right) = (6.4n)\ln\left(\frac{4 - 400m}{4 - 530m}\right) = 240\ ps$$

v_{SWO} when M_{EG} closes: $v_{DO} + v_O$ to 0 $\therefore$ $\Delta v_{SWO} = v_{DO} + v_O = 4.8\ V$

$$t_{V(C)} \approx \left(\frac{R_P}{v_{DD} - v_{TH(C)}}\right)\left[C_{OL}\Delta v_{SWO} + \left(\frac{0.5C_{CH}}{2}\right)v_{TH(C)}\right] = 1.6\ ns$$

$$P_{IV} \approx i_{L(LO)}\Delta v_{SWO}\left(\frac{t_{I(C)}}{3t_{SW}} + \frac{t_{V(C)}}{2t_{SW}}\right) = (470m)(4.8)\left[\frac{240p}{(3)1\mu} + \frac{1.6n}{(2)1\mu}\right] = 2.0\ mW$$

Boost $\therefore$ $d_{IN} = 1$ → $\sigma_{IV} = \dfrac{P_{IV}}{\left(i_{O(AVG)}/d_O\right)d_{IN}v_{IN}} = \dfrac{2.0m}{(250m/48\%)(1)(2)} = 0.2\%$

SPICE Simulation

```
* I-V Overlap Loss when MEG Closes
vdd vdd 0 dc=4
rp vdd vg 100
lx 0 vswo 10u
meg vswo vg 0 0 nmos1 w=50m l=250n
ddo vswo vo fdiode1   → τF = 0
vo vo 0 dc=4
.ic i(lx)=470m v(vg)=0
.lib lib.txt
.tran 5n
.end
```

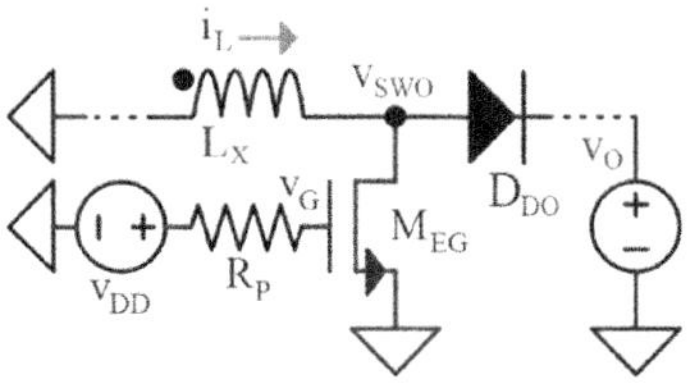

* Plot v_G, i_D, v_{SWO}

* Plot $i_{DS}v_{DS} \approx i_D v_{SWO}$

* Average $i_D v_{SWO}$ across 5 ns

* Average across t_{SW} → $\times\left(\dfrac{5\ ns}{t_{SW}}\right)$

* Caution: $i_D = i_{DS} + i_{GD} + i_{DB}$

B. Opening Switch

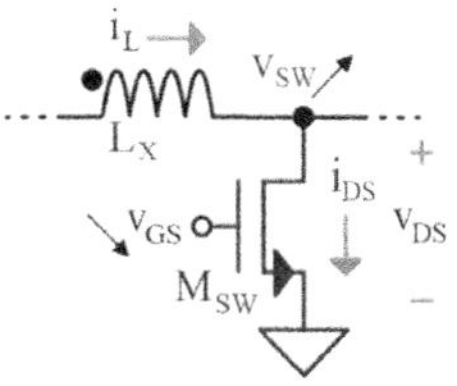

R_N discharges C_G to v_{GS} needed to sustain $i_L - i_N - i_{SW} \approx i_L$:

$$v_{TH(O)} \approx V_{T0} + v_{DS(SAT)}\Big|_{i_{L(O)}} \approx V_{T0} + \sqrt{\frac{2i_{L(O)}}{K'(W/L)}}$$

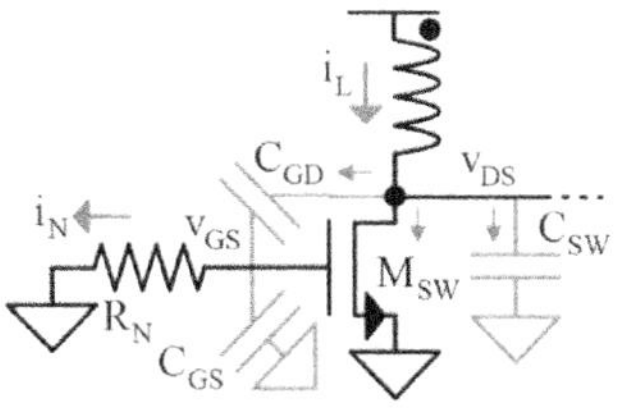

i_N slews C_{GD} to raise v_{DS} at v_{GS} needed to sustain i_L:

$$t_{V(O)} \approx \left(\frac{\Delta v_C}{i_N}\right)C_{GD} \approx \left(\frac{R_N}{v_{TH(O)}}\right)\left[C_{OL}\Delta v_{SW} + \left(\frac{C_{CH}}{4}\right)v_{TH(O)}\right]$$

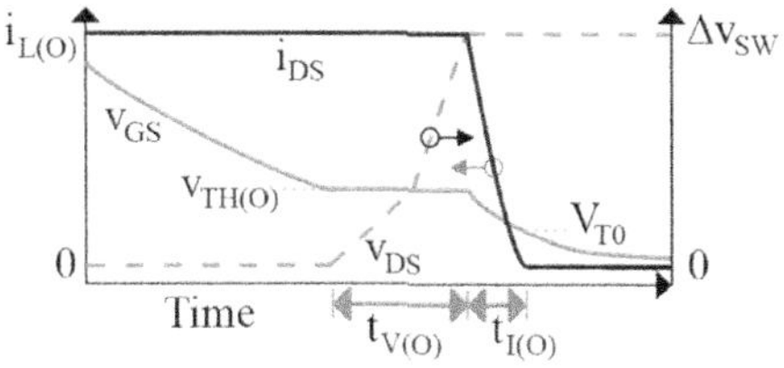

i_{DS} falls when v_{GS} falls from $v_{TH(O)}$ to V_{T0}:

$$t_{I(O)} \approx t_{T0} - t_{TH(O)} = \tau_{RC(O)}\ln\left[\frac{v_{DD} - \left(v_{DD} - v_{TH(O)}\right)}{v_{DD} - \left(v_{DD} - V_{T0}\right)}\right] \approx R_N\left[2C_{OL} + \left(\frac{2}{3}\right)C_{CH}\right]\ln\left(\frac{v_{TH(O)}}{V_{T0}}\right)$$

C. Reverse Recovery

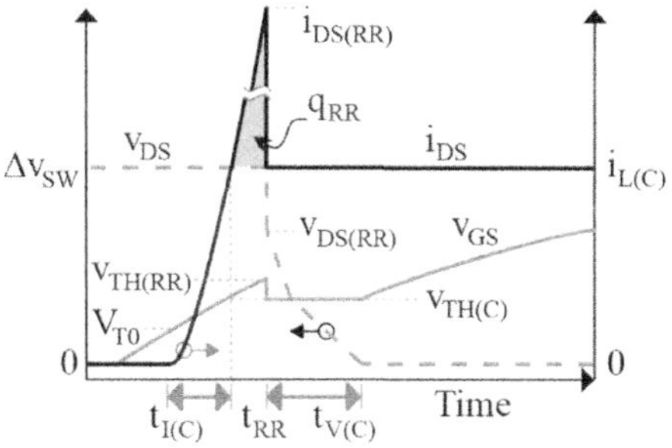

D_{DT} holds $q_{RR} = q_{DIF} = i_L\tau_T$ when conducting i_L.

i_{DS} pulls i_L & q_{RR} before discharging C_{GD} & C_{SW}.

Linear Approximation: $v_{DS(RR)} \approx 0 \;\rightarrow\; t_{V(C)} \approx 0$

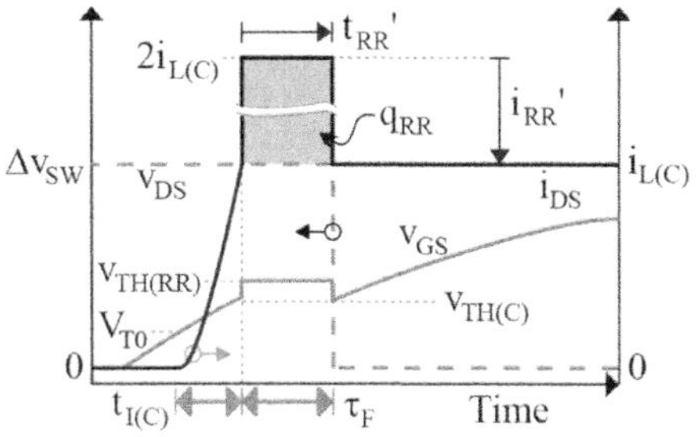

i_{DS} pulls q_{RR} with $i_{RR}' \approx i_{L(C)}$ across $t_{RR}' \approx \tau_T$.

$q_{RR} \approx i_{RR}'t_{RR}'$ → i_{RR}'↑ & t_{RR}'↓ shifts in P_{RR} cancel

$$\therefore\; P_{IV} = P_{IV(C)} + P_{IV(O)} \qquad P_{RR} \approx 2i_{L(C)}\Delta v_{SW}\left(\frac{\tau_T}{t_{SW}}\right)$$

$$\approx i_{L(C)}\Delta v_{SW}\left(\frac{t_{I(C)}}{3t_{SW}} + \frac{2\tau_T}{t_{SW}}\right) + i_{L(O)}\Delta v_{SW}\left(\frac{t_{I(O)}}{3t_{SW}} + \frac{t_{V(O)}}{2t_{SW}}\right)$$

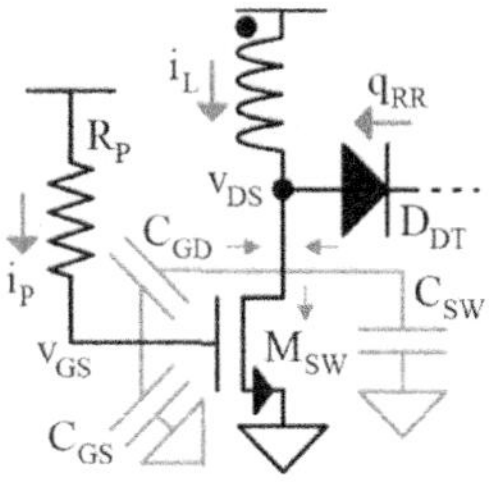

t_{DT} Conduction: Body diodes conduct & inject i_{SUB} $\therefore$ Generate substrate noise

Implicit MOS diodes conduct when $V_{T0} < 0.6–0.7$ V:

i_L discharges & charges C_{SWI} & C_{SWO} until M_{DG} & M_{DO} conduct i_L

$\therefore$ Lower i_{SUB} & $q_{RR} = q_{DIF} + q_{GS} \approx q_{GS}$ → Lower noise & lower P_{IV}

Add Schottky diodes:

Lower v_D, i_{SUB}, $q_{RR} \approx q_{DEP}$

→ Lower P_{DT}, noise, P_{IV}

More processing steps

Or off-chip components

→ Higher cost

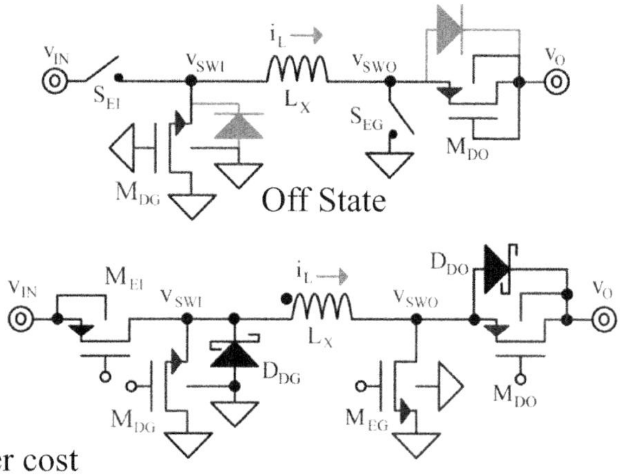

D. Soft Switching

Soft Switching ≡ Transition when i_{DS} or v_{DS} = Low $\therefore$ $P_{IV} \propto i_{DS}v_{DS}$ = Low

Examples: If $v_{DG} << v_{IN}$ → After t_{DT}, S_{DG} collapses v_{DG}

If $v_{DO} << v_O$ → After t_{DT}, S_{DO} collapses v_{DO}

} Drain Switches

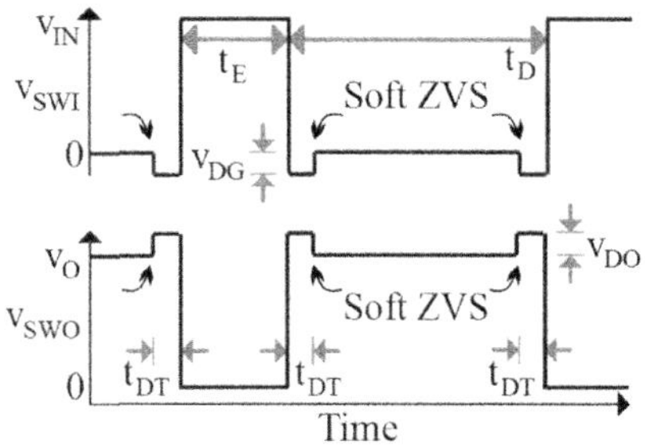

S_E's & L_X transition v_{SW}'s

$\therefore$ S_E's burn more P_{IV} than S_D's

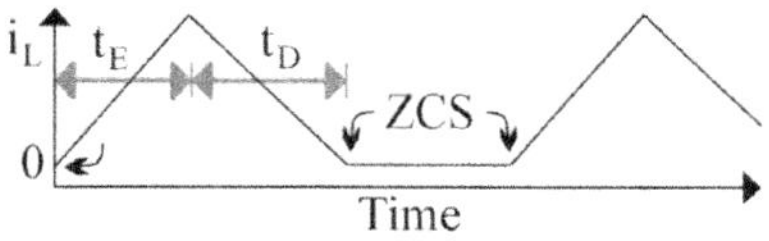

Zero-Voltage Switching (ZVS) ≡ Transition i_{DS} when $v_{DS} \approx 0$

Zero-Current Switching (ZCS) ≡ Transition v_{DS} when $i_{DS} \approx 0$

} $P_{IV} \approx 0$

Examples: DCM → Drain switches open when $i_L = 0$

Energize switches close when $i_L = 0$

4.7. Gate-Drive Loss

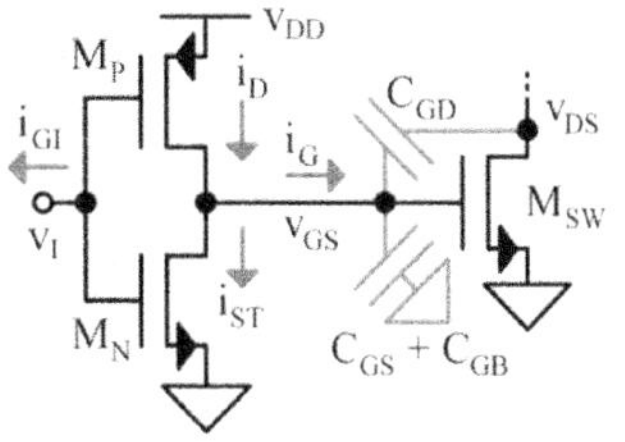

Gate Driver: M_P charges C_G M_N discharges C_G

v_{DD} supplies i_G & i_{ST} → $P_{DD} = P_G + P_{ST}$

C_G = High → v_I falls before v_{GS} rises

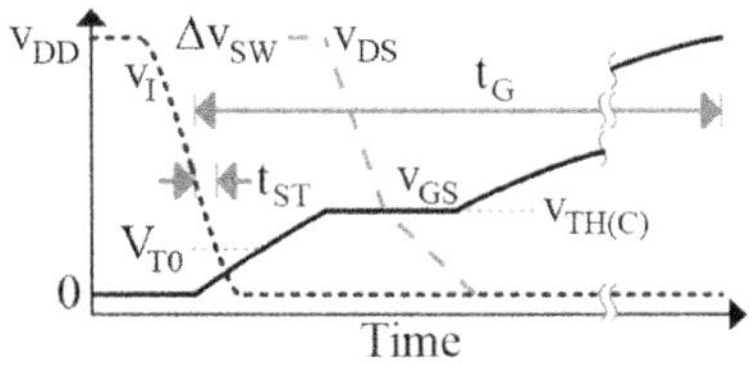

As v_I falls, $v_{GS} = v_{DSN}$ = Low

∴ $i_N = i_{ST}$ = Low

$$P_{DD} \approx P_G = v_{DD} i_{G(AVG)} = v_{DD}\left(\frac{q_G}{t_{SW}}\right)$$

Approximation: M_{SW} = Mostly in saturated inversion when $v_{SW} = v_{DS}$ collapses.

∴ All C_G's charge across v_{DD}

C_{OL} in C_{GD} charges across Δv_{SW}

$C_{GS} + C_{GB} + C_{GD}$ $\quad$ $C_{GD(SAT)}$

$$q_G \approx (2C_{OL} + C_{CH})v_{DD} + C_{OL}\Delta v_{SW}$$

Example: P_D & σ_D for M_{EG}'s driver in the ideal synchronous boost from previous example in CCM when W_{EG} = 50 mm, L_{EG} = 250 nm, L_{OL} = 30 nm.

Solution:

$d_{IN} = 1$, $d_E = 52\%$, $d_O = d_D = 48\%$, $\Delta i_L = 100$ mA (from previous example)

$\Delta v_{SWO} = 4.8$ V, $C_{OL} = 10$ pF, $C_{CH} = 66$ pF (from previous example)

$$q_G \approx C_{OL}\Delta v_{SW} + (2C_{OL} + C_{CH})v_{DD} = (10p)(4.8) + [2(10p) + 66p](4) = 390\text{ pC}$$

$$P_D \approx v_{DD}\left(\frac{q_G}{t_{SW}}\right) = 4\left(\frac{390p}{1\mu}\right) = 1.6\text{ mW}$$

$$\sigma_D = \frac{P_D}{\left(i_{O(AVG)}/d_O\right)d_{IN}v_{IN}} = \frac{1.6m}{(250m/49\%)(1)(2)} = 0.2\%$$

SPICE Simulation

```
* Ideal Boost with Dead Time, MEG, & MEG's Driver in CCM
vdeb deb 0 dc=4 pulse 4 0 0 1n 1n 515n 1u
vdo do 0 dc=1 pulse 0 1 565n 1n 1n 385n 1u
vdd vdd 0 dc=4
mp vg deb vdd vdd pmos1 w=26u l=250n
mn vg deb 0 0 nmos1 w=14u l=250n
vin vin 0 dc=2
lx vin vswo 10u
meg vswo vg 0 0 nmos1 w=50m l=250n
sdo vswo vo do 0 sw1v
ddo vswo vo fdiode1
vo vo 0 dc=4
```

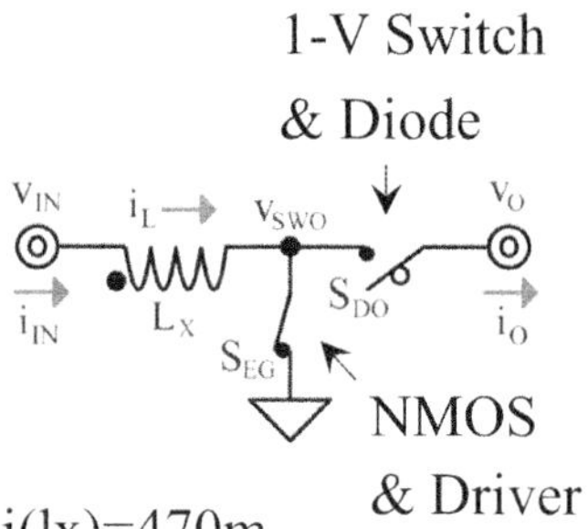


```
.ic i(lx)=470m
.lib lib.txt
.tran 1u
.end
```

* Plot v_G, v_{SWO}, i_L, $i_{DD}v_{DD}$

* Average $i_{DD}v_{DD}$

* Caution: $i_{DD} = i_G + i_{ST} \approx i_G$

4.8. Leaks

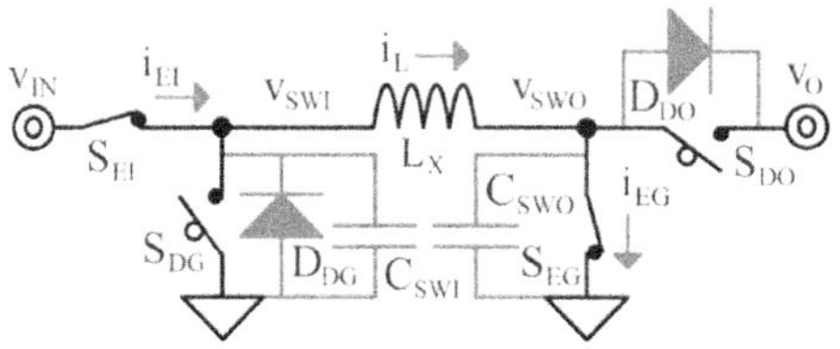

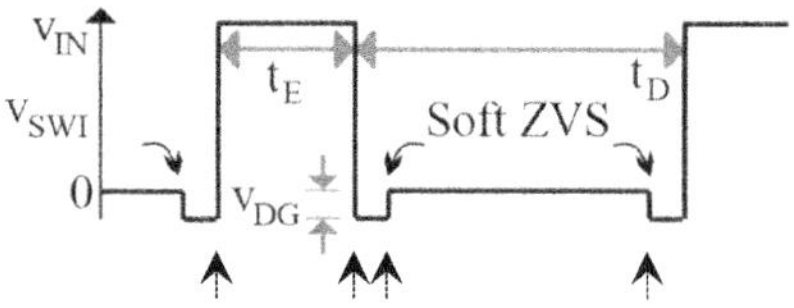

C_{SWI}:

v_{SWI} rises from $-v_{DG}$ to v_{IN} → v_{IN} supplies $v_{IN}q_{SWI} = v_{IN}C_{SWI}(v_{IN} + v_{DG})$

→ C_{SWI} loses $0.5C_{SWI}v_{DG}^2$ & receives $0.5C_{SWI}v_{IN}^2$

L_X collapses v_{SWI} to $-v_{DG}$ → L_X helps v_O recover $0.5C_{SWI}v_{IN}^2$

L_X charges C_{SWI} to $-v_{DG}$ → Loses $0.5C_{SWI}v_{DG}^2$

M_{DG} discharges C_{SWI} to 0 V → C_{SWI} loses $0.5C_{SWI}v_{DG}^2$

L_X charges C_{SWI} to $-v_{DG}$ → L_X loses $0.5C_{SWI}v_{DG}^2$

$$\therefore\ v_{IN}, L_X, v_O \text{ lose } P_{SWI} = E_{SWI}f_{SW} = \left(\frac{C_{SWI}}{t_{SW}}\right)\left(0.5v_{IN}^2 + v_{IN}v_{DG} + v_{DG}^2\right)$$

C_{SWO}:

M_{EG} drains C_{SWO} to ground → C_{SWO} loses $0.5C_{SWO}(v_O + v_{DO})^2$

L_X charges C_{SWO} to $v_O + v_{DO}$ → Loses $0.5C_{SWO}(v_O + v_{DO})^2$

M_{DO} discharges C_{SWO} to v_O → v_O recovers $v_O q_{SWO} = v_O C_{SWO} v_{DO}$

L_X charges C_{SWO} to $v_O + v_{DO}$ → Loses $0.5C_{SWO}[(v_O + v_{DO})^2 - v_O^2]$

$$\therefore \quad v_{IN}, L_X, v_O \text{ lose } P_{SWO} = E_{SWO} f_{SW} = \left(\frac{C_{SWO}}{t_{SW}}\right)\left(0.5 v_O^2 + v_{DO}^2 + v_O v_{DO}\right)$$

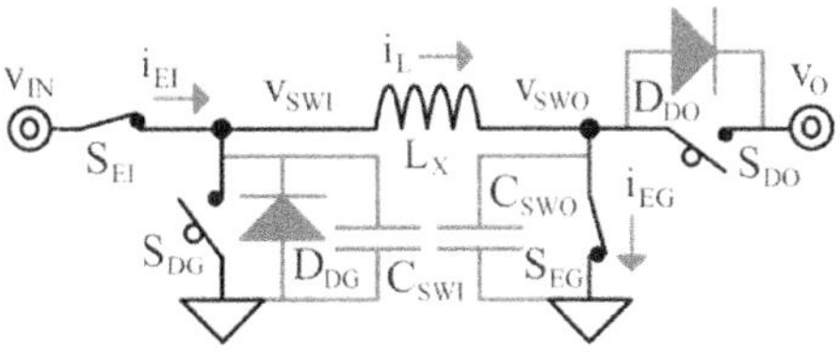

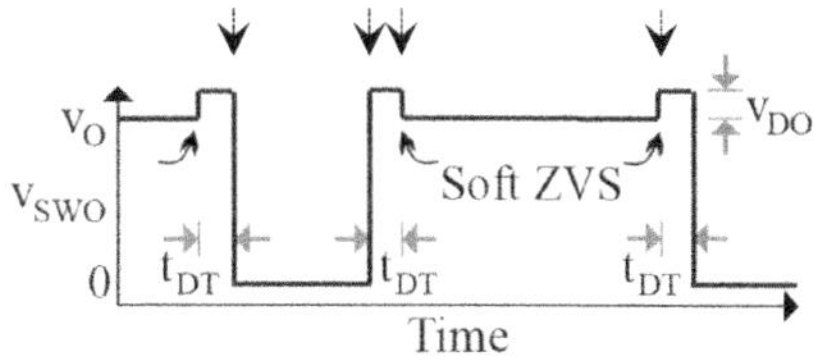

Cut-Off Power

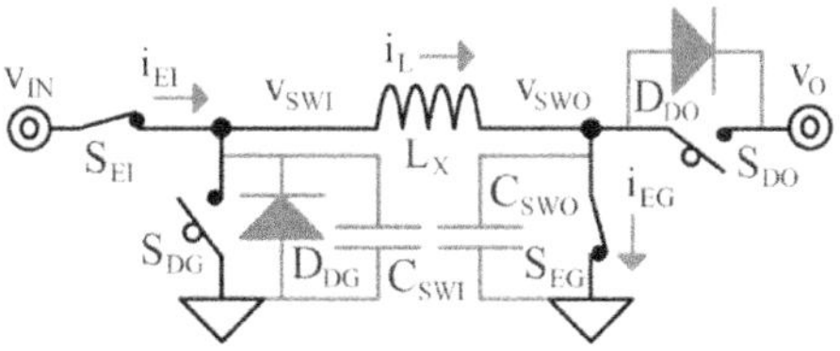

Large Transistors:

Leak i_{DS} & $i_{DB} \approx I_S$ → $i_{OFF} \propto W_{CH}, T_J$ → R_{OFF} falls with rising W_{CH}, T_J

Of every transistor pair, one is always off:

When $R_{I/O}$'s ≈ Similar

$$P_{OFF} \approx \left(\frac{v_{IN}^2}{R_{DG(OFF)}} + \frac{v_O^2}{R_{DO(OFF)}}\right)\left(\frac{t_E}{t_{SW}}\right) + \left(\frac{v_{IN}^2}{R_{EI(OFF)}} + \frac{v_O^2}{R_{EG(OFF)}}\right)\left(\frac{t_D}{t_{SW}}\right) \approx \frac{v_{IN}^2}{R_{I(OFF)}} + \frac{v_O^2}{R_{O(OFF)}}$$

Measured Example:

$W_{CH} = 30$ mm, $L_{OX} = 180$ nm → Leaks 80 nA with 1.8 V at 25°C

10× higher at 125°C → $R_{OFF(SQ)} = 0.38$–3.8 TΩ per L/W Square

4.9. Design: A. Power Switch

$P_R \propto R_{CH} \propto L_{CH}$ $\quad P_G \propto C_{CH} \propto L_{CH}$ $\quad\therefore\quad$ Use L_{MIN} needed to sustain $v_{IN/O}$

$$P_R = i_{R(RMS)}{}^2 R_{CH} = \frac{k_R}{W_{CH}} \qquad P_G = E_G f_{SW} = v_{DD} q_G f_{SW} = v_{DD} C_G \Delta v_G f_{SW} = k_G W_{CH}$$

$\therefore$ Raise W_{CH} until additional P_G losses cancel P_R savings → Optimal W_{CH}'

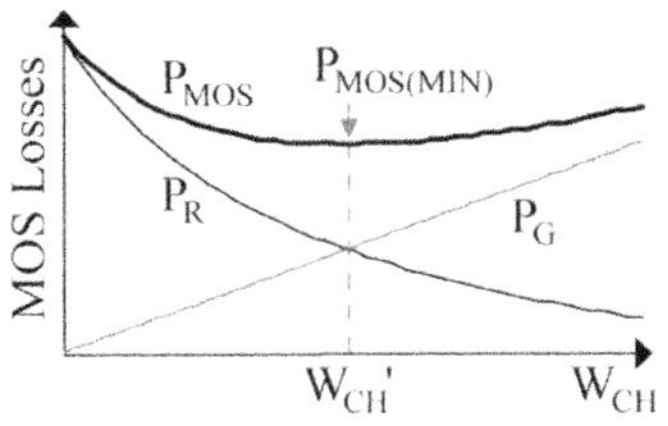

$$\left.\frac{\partial P_{MOS}}{\partial W_{CH}}\right|_{W_{CH}'} = \left.\frac{\partial P_R}{\partial W_{CH}}\right|_{W_{CH}'} + \left.\frac{\partial P_G}{\partial W_{CH}}\right|_{W_{CH}'} = -\frac{k_R}{W_{CH}'^2} + k_G = 0$$

$$\therefore \quad W_{CH}' = \sqrt{\frac{k_R}{k_G}} \qquad P_R\big|_{W_{CH}'} = P_G\big|_{W_{CH}'} = \sqrt{k_R k_G}$$

$$P_{MOS(MIN)} = P_{MOS}\big|_{W_{CH}'} = P_R\big|_{W_{CH}'} + P_G\big|_{W_{CH}'} = 2P_R\big|_{W_{CH}'} = 2P_G\big|_{W_{CH}'} = 2\sqrt{k_R k_G}$$

Example: M_{DG}'s optimal W_{DG}, L_{DG}, P_{MOS}, & σ_{MOS} in an ideal sync. buck–boost when $v_{DD} = v_{IN} = 2$ V, $v_O = 4$ V, D_{DG} & D_{DO} drop 400 mV, $i_{O(AVG)} = 250$ mA, $L_X = 10$ μH, $t_{SW} = 1$ μs, $t_{DT} = 50$ ns, $L_{MIN} = 250$ nm, $L_{OL} = 30$ nm, $V_{TN0} = 400$ mV, $K_N' = 200$ μA/V^2, $C_{OX}'' = 6.9$ fF/μm^2.

Solution:

$$d_{IN} = d_E \approx \frac{v_O}{v_{IN} + v_O} + \left(\frac{v_{DO} + v_{DG}}{v_{IN} + v_O}\right)\left(\frac{2t_{DT}}{t_{SW}}\right) = \frac{4}{2+4} + \left(\frac{400m + 400m}{2+4}\right)\left[\frac{2(50n)}{1\mu}\right] = 68\%$$

$$\therefore \quad d_O = d_D = 1 - d_E \approx 32\%$$

$$\Delta i_L = \left(\frac{v_E}{L_X}\right) d_E t_{SW} \approx \left(\frac{2}{10\mu}\right)(68\%)(1\mu) = 140 \text{ mA}$$

$L_{DG} \equiv L_{MIN} = 250$ nm $\;\therefore\;$ $L_{CH} = L_{DG} - 2L_{OL} = 250n - 2(30n) = 190$ nm

$$R_{DG} \approx \left(\frac{L_{CH}}{W_{CH}}\right)\left[\frac{1}{K_N'(v_{DD} - V_{TN0})}\right] = \left(\frac{190n}{W_{CH}}\right)\left[\frac{1}{(200\mu)(2-400m)}\right] = \frac{590\mu}{W_{CH}}$$

$$P_R = \left[\left(\frac{i_{O(AVG)}}{d_O}\right)^2 + \left(\frac{0.5\Delta i_L}{\sqrt{3}}\right)^2\right] R_{DG}\left(d_D - \frac{2t_{DT}}{t_{SW}}\right)$$

$$\approx \left\{\left(\frac{250m}{32\%}\right)^2 + \left[\frac{0.5(140m)}{\sqrt{3}}\right]^2\right\}\left(\frac{590\mu}{W_{CH}}\right)\left[32\% - \frac{2(50n)}{1\mu}\right] = \frac{79\mu}{W_{CH}}$$

$$C_{OL} = C_{OX}''W_{CH}L_{OL} = (6.9m)W_{CH}(30n) = (210p)W_{CH}$$

$$C_{CH} = C_{OX}''W_{CH}L_{CH} = (6.9m)W_{CH}(190n) = (1.3n)W_{CH}$$

v_{SWI} when M_{EI} closes: $-v_{DG}$ to v_{IN} $\quad\therefore\quad \Delta v_{SWI} = v_{DG} + v_{IN} = 400m + 2 = 2.4\ V$

$$q_G \approx C_{OL}\Delta v_{SWI} + (2C_{OL} + C_{CH})v_{DD} = \{(210p)(2.4) + [2(210p) + 1.3n](2)\}W_{CH}$$

$$= (3.9n)W_{CH}$$

$$P_G = v_{DD}\left(\frac{q_G}{t_{SW}}\right) \approx (2)\left[\frac{(3.9n)W_{CH}}{1\mu}\right] = (7.8m)W_{CH}$$

$$\therefore \quad W_{DG} \equiv W_{CH}' = \sqrt{\frac{k_R}{k_G}} = \sqrt{\frac{79\mu}{7.8m}} = 100\ mm$$

$$R_{DG} \approx \frac{590\mu}{W_{DG}} = 5.9\ m\Omega$$

$$P_{MOS} \approx 2\sqrt{k_R k_G} = 2\sqrt{(79\mu)(7.8m)} = 1.6\ mW$$

$$\sigma_{MOS} = \frac{P_{MOS}}{\left(i_{O(AVG)}/d_O\right)d_{IN}v_{IN}} = \frac{1.6m}{(250m/32\%)(68\%)(2)} = 0.2\%$$

In Practice:

On-chip & PCB

metal traces & vias

add resistance.

→ Higher R_{DG} & P_R

B. Gate Driver

Balance P_{IV}'s across t_{SW}

→ $t_{V(C)} \equiv t_{V(O)}$ ∴ →

$$\frac{R_P}{R_N} = \left(\frac{W_N}{W_P}\right)\left(\frac{L_P}{L_N}\right)\left(\frac{K_N'}{K_P'}\right)\left[\frac{v_{DD} - V_{TN0} - 0.5v_{TH(O)}}{v_{DD} - |V_{TP0}| - 0.5\left(v_{DD} - v_{TH(C)}\right)}\right]$$

$$\equiv \left(\frac{v_{DD} - v_{TH(C)}}{v_{TH(O)}}\right)\left(\frac{C_{OL}\Delta v_{SW} + 0.25C_{CH}v_{TH(O)}}{C_{OL}\Delta v_{SW} + 0.25C_{CH}v_{TH(C)}}\right)$$

Without q_{RR}

↓

$$P_{IV} = P_{I(C)}\left(\frac{t_{I(C)}}{t_{SW}}\right) + P_{V(C)}\left(\frac{t_{V(C)}}{t_{SW}}\right) + P_{I(O)}\left(\frac{t_{I(O)}}{t_{SW}}\right) + P_{V(O)}\left(\frac{t_{V(O)}}{t_{SW}}\right)$$

$$\approx \left(i_{L(LO)} + i_{L(HI)}\right)\Delta v_{SW}\left(\frac{t_I}{3t_{SW}} + \frac{t_V}{2t_{SW}}\right)$$

$$= 2\left(\frac{i_{O(AVG)}}{d_O}\right)\Delta v_{SW}\left(\frac{t_I}{3t_{SW}} + \frac{t_V}{2t_{SW}}\right) = \frac{k_{IV}}{W_N}$$

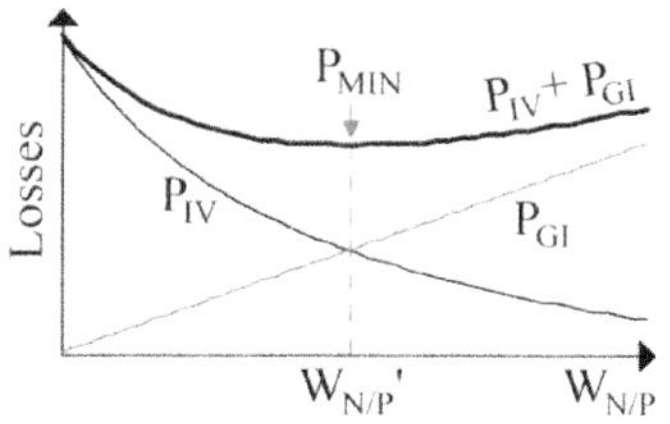

$$P_{GI} = v_{DD}\left(\frac{q_{GI}}{t_{SW}}\right) = v_{DD}\left(\frac{q_{GN} + q_{GP}}{t_{SW}}\right) = k_{GI}W_N$$

∴ Raise W_N until additional P_{GI} losses cancel P_{IV} savings → Optimal $W_N' = \sqrt{\frac{k_{IV}}{k_{GI}}}$

C. Operation: Switch Configuration

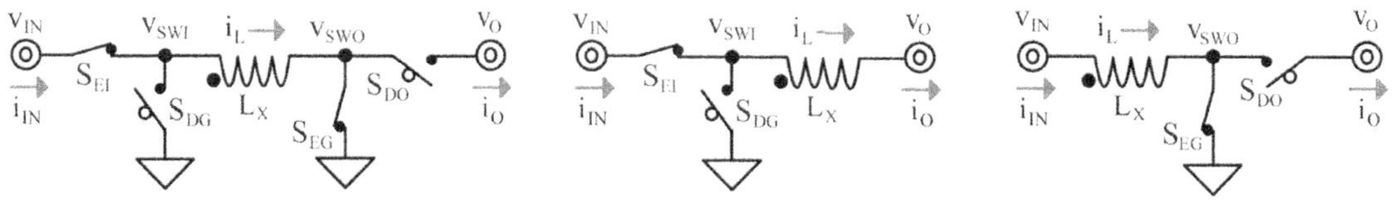

Ideal Buck–Boost: $P_{IN} \approx P_L \approx P_O$ ∴ Use buck–boost only when necessary

Direct Transfers: Buck: v_{IN} supplies v_O as L_X energizes
Boost: v_{IN} supplies v_O as L_X drains
} $P_{IN} \approx P_O > P_L$

∴ Same P_O with lower $i_{L(AVG)}$ → Lower $P_R \Leftrightarrow P_G$
fewer switches → Smaller A_{Si}

Optimal Buck–Boost: When bucking → Keep S_{EG} open & S_{DO} closed
When boosting → Keep S_{EI} closed & S_{DG} open

∴ Lower $i_{L(AVG)}$ & Lower P_R & P_G

Discontinuous Conduction

Optimize W_{CH} across t_C → E_L packet

$$P_{R(C)} \approx i_{E/D(RMS)}{}^2 R_{E/D}\left(\frac{t_{E/D}}{t_C}\right) = \left(\frac{i_{L(PK)}}{\sqrt{3}}\right)^2 R_{E/D}\left(\frac{t_{E/D}}{t_C}\right) = \frac{k_{RC}}{W_{CH}}$$

$$W_{E/D} \equiv W_{CH}' = \sqrt{\frac{k_{RC}}{k_{GC}}}$$

$$P_{G(C)} = v_{DD} i_{G(AVG)} = v_{DD}\left(\frac{q_G}{t_C}\right) = k_{GC} W_{CH}$$

$$P_{MOS} = P_{MOS(C)}\left(\frac{t_C}{t_{SW}}\right) = 2\left(\sqrt{k_{RC} k_{GC}}\right) t_C f_{SW}$$

$$P_O \approx \frac{E_L}{t_{SW}} + P_{E/D}\left(\frac{t_{E/D}}{t_{SW}}\right)$$

Direct Transfer (under the $P_{E/D}$ term); d_O or $d_{IN} = 1$ (under the $\frac{i_{L(PK)}}{2}$ term):

$$\approx \left[\left(\frac{1}{2}\right) L_X i_{L(PK)}{}^2 + \left(\frac{i_{L(PK)}}{2}\right) v_{O/IN} t_{E/D}\right] f_{SW}$$

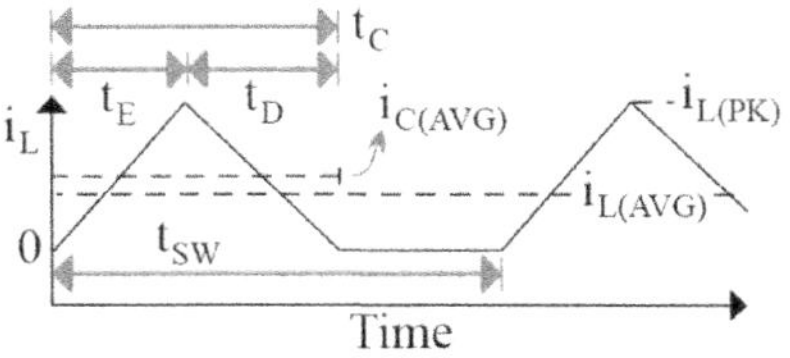

$$\eta_C = \frac{P_O}{P_{IN}} \approx \frac{P_O}{P_O + P_{MOS} + P_{DT} + P_{IV} + P_{GI} + P_{SW}} \propto \frac{f_{SW}}{f_{SW}} \neq f(P_O)$$

∴ Max η_C at $i_{L(PK)}$ with W_{CH}'

$\eta_C \neq f(P_O)$ when adjusting f_{SW} to supply P_O

→ η_C = Optimally high across P_O in DCM with frequency modulation (FM)

D. Power-Conversion Efficiency: Discontinuous Conduction

Frequency Modulation: $i_{L(PK)}$ & t_C = Constant → i_O scales with f_{SW}

$$\sigma_{DCM0} \approx \frac{P_{CNTRL} + P_{OFF}}{P_{IN}} \propto \frac{i_O{}^0}{i_O{}^1} \propto \frac{1}{i_O}$$ → Falls with higher i_O ∴ η_C Rises

$$\sigma_{DCM1} \approx \frac{P_R + P_{DT} + P_{IV} + P_G + P_{GI} + P_{SW}}{P_{IN}} \propto \frac{i_O{}^1}{i_O{}^1} \neq f(i_O)$$ → Flat ∴ η_C Flattens

Peak Modulation: f_{SW} = Constant → $i_{L(PK)}$ & t_C scale with $\sqrt{i_O}$

$$\sigma_{DCM0} \approx \frac{P_G + P_{GI} + P_{SW} + P_{CNTRL} + P_{OFF}}{P_{IN}} \propto \frac{i_O{}^0}{i_O{}^1} = \frac{1}{i_O}$$ → Falls with higher i_O ∴ η_C Rises

$$\sigma_{DCM0.5} \approx \frac{P_{DT} + P_{IV}}{P_{IN}} \propto \frac{i_O{}^{0.5}}{i_O{}^1} = \frac{1}{\sqrt{i_O}}$$ → Falls ∴ Slow η_C Rise

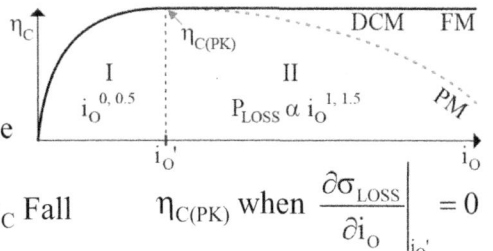

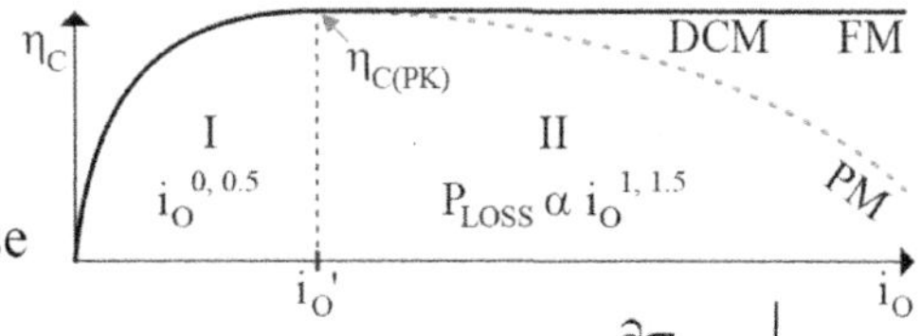

$$\sigma_{DCM1.5} \approx \frac{P_R}{P_{IN}} \propto \frac{i_O{}^{1.5}}{i_O{}^1} = \sqrt{i_O}$$ → Rises ∴ Slow η_C Fall

$\eta_{C(PK)}$ when $\left.\frac{\partial \sigma_{LOSS}}{\partial i_O}\right|_{i_O'} = 0$

Continuous Conduction

$$\sigma_{CCM0} \approx \frac{P_{R(AC)} + P_G + P_{GI} + P_{SW} + P_{CNTRL} + P_{OFF}}{P_{IN}} = \frac{k_{C0}}{i_O} \rightarrow \text{Falls with } i_O \;\therefore\; \eta_C \text{ Rises}$$

$$\sigma_{CCM1} \approx \frac{P_{DT} + P_{IV}}{P_{IN}} = k_{C1}\left(\frac{i_O}{i_O}\right) = k_{C1} \rightarrow \text{Flat} \;\therefore\; \eta_C \text{ Flattens}$$

$$\sigma_{CCM2} \approx \frac{P_{R(DC)}}{P_{IN}} = k_{C2}\left(\frac{i_O^{\;2}}{i_O}\right) = k_{C2} i_O \rightarrow \text{Rises} \;\therefore\; \eta_C \text{ Falls}$$

η_C peaks when:

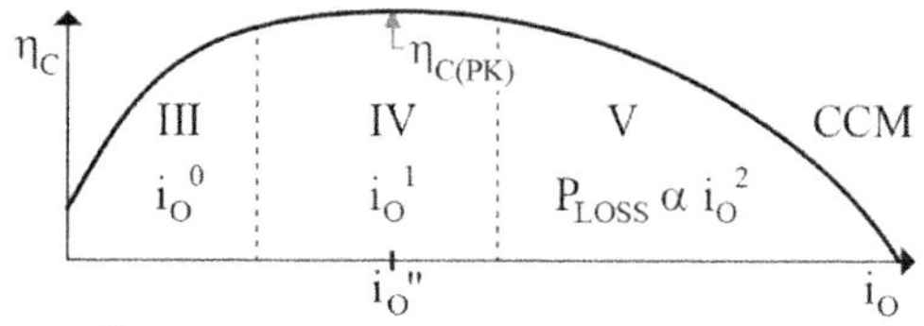

$$\left.\frac{\partial \sigma_{CCM0}}{\partial i_O} + \frac{\partial \sigma_{CCM1}}{\partial i_O} + \frac{\partial \sigma_{CCM2}}{\partial i_O}\right|_{i_O"} = -\frac{k_{C0}}{i_O"^2} + 0 + k_{C2} = 0$$

$$\therefore \quad i_O" = \sqrt{\frac{k_{C0}}{k_{C2}}} \qquad \sigma_{CCM0}\big|_{i_O"} = \sigma_{CCM2}\big|_{i_O"} = \sqrt{k_{C0} k_{C2}} \qquad \eta_{C(PK)} = 1 - \sigma_{CCM1} - 2\sqrt{k_{C0} k_{C2}}$$

Chapter 5. Frequency Response

5.1. Two-Port Models

5.2. LC Primitives

5.3. Bypass Capacitors

5.4. Voltage-Sourced LC

5.5. Switched Inductor

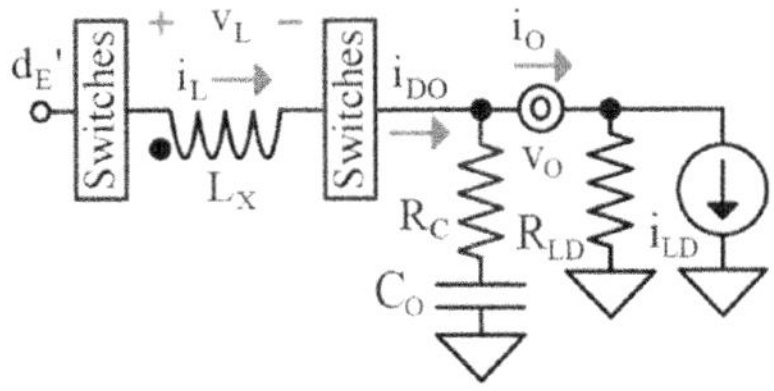

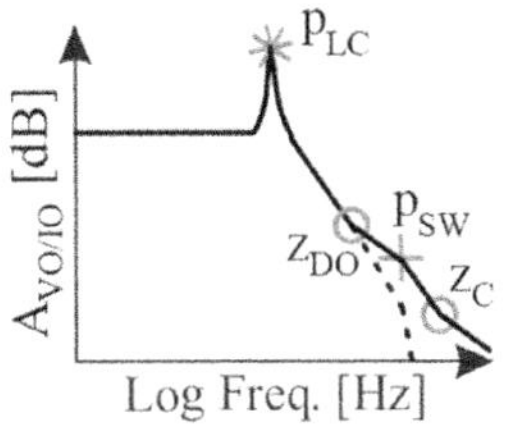

5.1. Two-Port Models: Primitives

Purpose: Predict loaded response with simple model

Principle: Orthogonal (mutually independent) components

Extraction: Nullify other components

Thévenin:

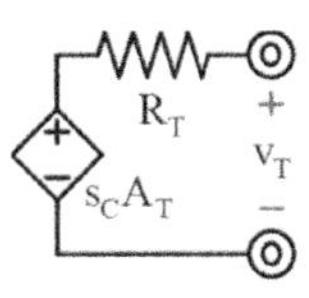

Derive A_T w/o R_T → $v_R \equiv 0$ ∴ $i_T \equiv 0$ → Remove load

Derive R_T w/o A_T ∴ $s_C \equiv 0$

Norton:

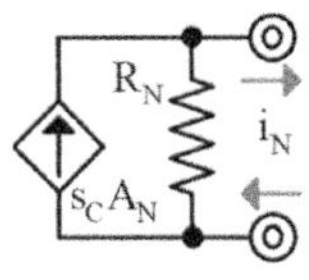

Derive A_N w/o R_N → $i_R \equiv 0$ ∴ $v_N \equiv 0$ → Short output

Derive R_N w/o A_N ∴ $s_C \equiv 0$

Bidirectional Example: Reverse Hybrid

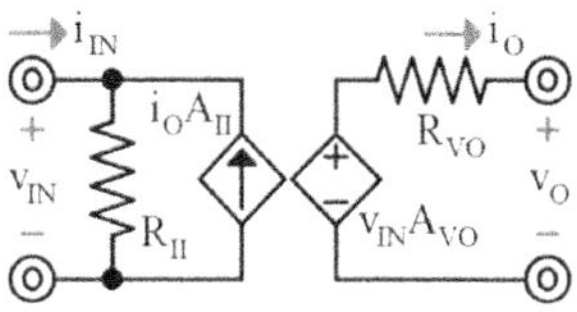

Forward Example: i-Sourced v_O

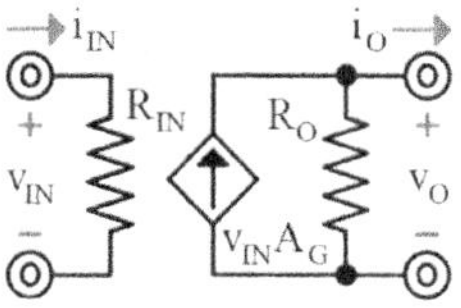

Feedback Model: Derive R_{II} when $i_O \equiv 0$ ∴ Remove load

Derive A_{II} when $i_{RI} \equiv 0$ ∴ $v_{IN} \equiv 0$ → Short input

Derive A_{VO} when $v_{RO} \equiv 0$ ∴ $i_O \equiv 0$ → Remove load

Derive R_{VO} when $v_{IN} \equiv 0$ ∴ Short input

Note: $A_{I/O}$ = Feedback/forward translations $R_{I/O}$ test conditions disable $A_{I/O}$

→ $A_{I/O}$ & $R_{I/O}$ = Open-loop parameters that model closed-loop behavior

Forward Model: Derive R_{IN} as is → No redundancies to consider

Derive A_G when $i_{RO} \equiv 0$ ∴ $v_O \equiv 0$ → Short output

Derive R_O when $v_{IN} \equiv 0$ ∴ Short input

5.2. LC Primitives: A. Impedances

Capacitor: $$Z_C = \frac{1}{sC_X} = \frac{1}{i\omega_O C_X} = \frac{1}{i(2\pi f_O)C_X} \propto \frac{1}{f_O}$$

$i^2 = -1$ ←

Angular Freq.: Radians per second

Frequency: Cycles per second

2π Radians per Cycle

∴ C_X opens at low f_O & shorts with f_O

Approx.: $\|$ R_X fades when $Z_C < R_X$ C_X shorts when ⊕ R_X limits i_C

Inductor: $Z_L = sL_X = i\omega_O L_X = i(2\pi f_O)L_X \propto f_O$

∴ L_X shorts at low f_O & opens with f_O

Approx.: ⊕ R_X fades when $Z_L > R_X$ L_X opens when $\|$ R_X limits v_L

B. Shunt Capacitor: Response

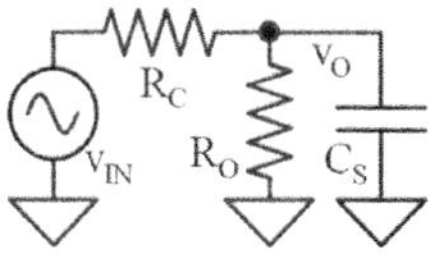

Concept: C_S shunts i_S away from R_O with f_O → Reduces v_O

C_S opens at low f_O: $A_{V0} = \dfrac{R_O}{R_C + R_O}$ ↓ Zero/Low-f_O Gain

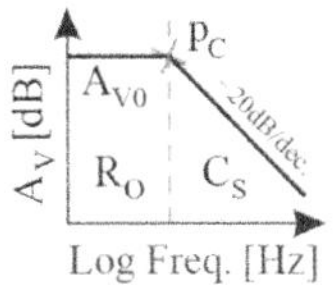

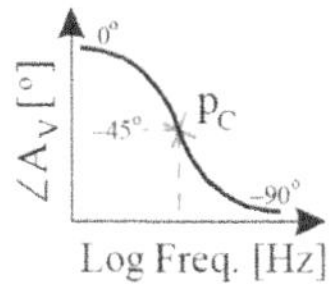

C_S shunts parallel R: $\left.\dfrac{1}{sC_S}\right|_{f_O \geq \frac{1}{2\pi(R_O \| R_C)C_S} \equiv p_C} \leq R_O \| R_C$

R_O fades past p_C ← Capacitor Pole ←

$A_V\big|_{f_O > p_C} \approx \dfrac{Z_S}{R_C} \propto \dfrac{1}{f_O}$ ↓ Falls 10× with 10× rise in f_O ↓ ↓ –20 dB per decade

Gain: $A_V \equiv \dfrac{v_O}{v_{IN}} = \dfrac{R_O \| Z_S}{R_C + (R_O \| Z_S)} = \dfrac{A_{V0}}{1 + s/2\pi p_C}$

C_S delays v_{IN}–v_O sinusoids: Phase shifts up to –90° past p_C → $\angle A_V = -\tan^{-1}\left(\dfrac{f_O}{p_C}\right)$

Current-Limit Resistor

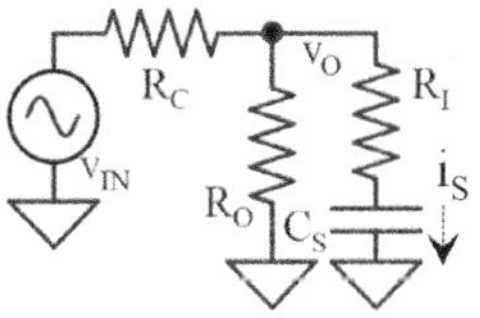

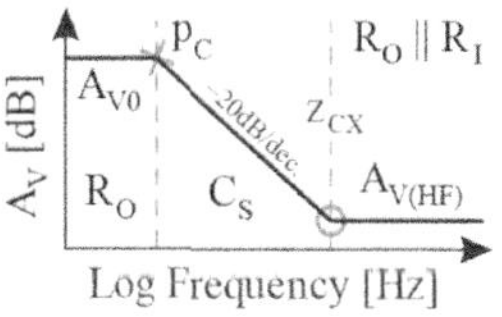

C_S shunts i_S → p_C

R_I limits i_S → Eliminates C_S effects → Reverses p_C

$A_{V0} = \dfrac{R_O}{R_C + R_O}$

C_S shunts parallel R: $\left.\dfrac{1}{sC_S}\right|_{f_O \geq \frac{1}{2\pi[R_I + (R_O \| R_C)]C_S} \equiv p_C} \leq R_I + (R_O \| R_C)$

$p_C < z_{CX}$

C_S shorts: $\left.\dfrac{1}{sC_S}\right|_{f_O \geq \frac{1}{2\pi R_I C_S} \equiv z_{CX}} \leq R_I$ → $A_V\big|_{f_O > z_{CX}} \approx \dfrac{R_O \| R_I}{R_C + (R_O \| R_I)}$ ↓ $A_{V(HF)}$ $\angle A_V \geq -90°$

(relative to R_I) ↳ Reversal Zero

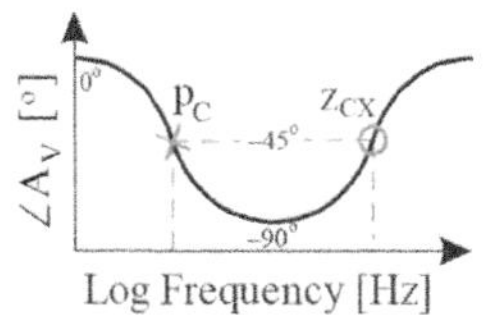

Gain: $A_V \equiv \dfrac{v_O}{v_{IN}} = \dfrac{R_O \| (Z_S + R_I)}{R_C + [R_O \| (Z_S + R_I)]} = A_{V0}\left(\dfrac{1 + s/2\pi z_{CX}}{1 + s/2\pi p_C}\right)$

Phase: z_{CX} recovers up to 90° → $\angle A_V = -\tan^{-1}\left(\dfrac{f_O}{p_C}\right) + \tan^{-1}\left(\dfrac{f_O}{z_{CX}}\right)$

C. Couple Inductor: Response

Concept: L_C impedes i_C with f_O → Reduces i_C into v_O → Reduces A_V to v_O

L_C shorts at low f_O: $A_{V0} \approx \dfrac{R_O}{0+R_O} = 1$

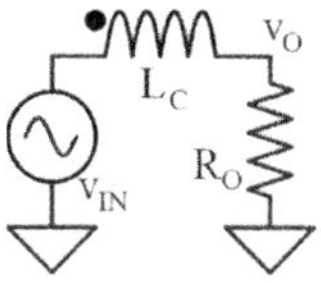

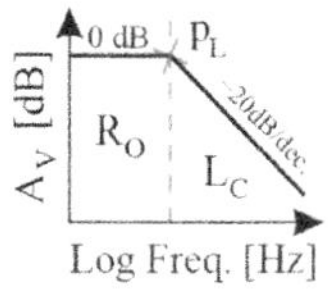

L_C overcomes series R: $sL_C\big|_{f_O \geq \frac{R_O}{2\pi L_C} \equiv p_L} \geq R_O$

Inductor Pole

$A_V\big|_{f_O > p_L} \approx \dfrac{R_O}{Z_C} = \dfrac{R_O}{sL_C} \propto \dfrac{1}{f_O}$

Falls 20 dB per decade

Gain: $A_V \equiv \dfrac{v_O}{v_{IN}} = \dfrac{R_O}{Z_C + R_O} = \dfrac{R_O}{sL_C + R_O} = \dfrac{A_{V0}}{1+s/2\pi p_L}$

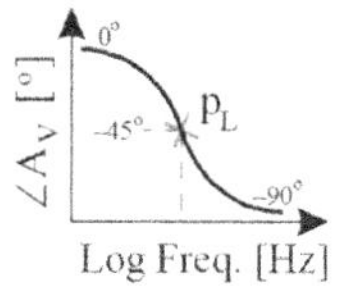

L_C delays v_{IN}–i_C sinusoids: Phase shifts up to –90° past p_L → $\angle A_V = -\tan^{-1}\left(\dfrac{f_O}{p_L}\right)$

Voltage-Limit Resistor

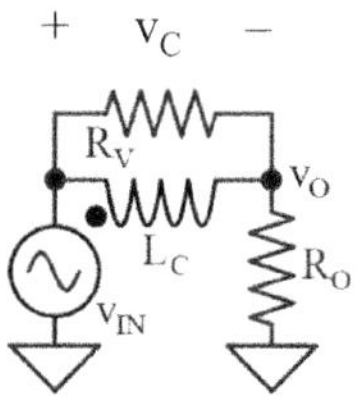

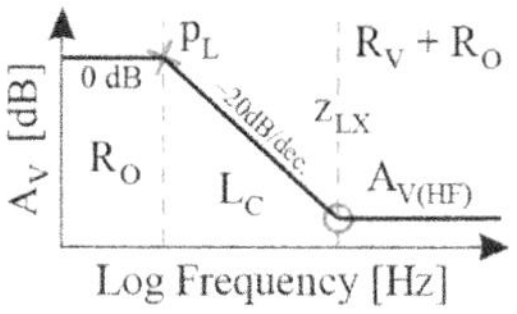

L_C impedes i_C → p_L

R_V limits v_C → Eliminates L_C effects → Reverses p_L

$A_{V0} \approx \dfrac{R_O}{0+R_O} = 1$

$\angle A_V \geq -90°$

L_C overcomes series R: $sL_C\big|_{f_O \geq \frac{R_O \| R_V}{2\pi L_C} \equiv p_L} \geq R_O \| R_V$

Reversal Zero

$A_{V(HF)}$

L_C opens when L_C surpasses R_V: $sL_C\big|_{f_O \geq \frac{R_V}{2\pi L_C} \equiv z_{LX}} \geq R_V$ → $A_V\big|_{f_O > z_{LX}} \approx \dfrac{R_O}{R_V + R_O}$

Gain: $A_V \equiv \dfrac{v_O}{v_{IN}} = \dfrac{R_O}{(Z_C \| R_V) + R_O} = A_{V0}\left(\dfrac{1+s/2\pi z_{LX}}{1+s/2\pi p_L}\right)$

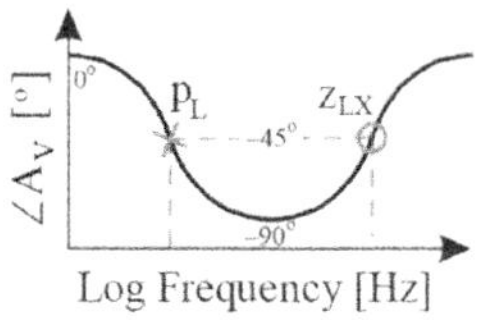

Phase: z_{LX} recovers up to 90° → $\angle A_V = -\tan^{-1}\left(\dfrac{f_O}{p_L}\right) + \tan^{-1}\left(\dfrac{f_O}{z_{LX}}\right)$ $\qquad p_L < z_{LX}$

5.3. Bypass Capacitors: A. Bypassed Resistor

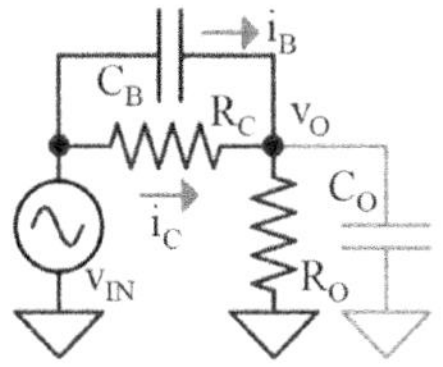

Concept: C_B feeds i_B → Increasing i_B raises v_O

C_B opens at low f_O: $A_{V0} = \frac{R_O}{R_C + R_O}$

C_B bypasses R_C: $\left.\frac{1}{sC_B}\right|_{f_O \geq \frac{1}{2\pi R_C C_B} \equiv z_C} \leq R_C$

→ Capacitor Zero

→ R_C fades

C_B shunts parallel R: $\left.\frac{1}{sC_B}\right|_{f_O \geq \frac{1}{2\pi (R_C \| R_O) C_B} \equiv p_{CX}} \leq R_C \| R_O$

→ Reversal Pole

→ C_B shorts

$$\to A_V\Big|_{f_O > p_{CX}} \approx \frac{R_O}{0 + R_O} = 1 \quad (\uparrow A_{V(HF)})$$

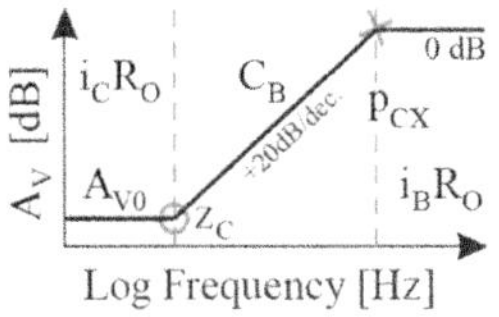

$$\text{Gain: } A_V \equiv \frac{v_O}{v_{IN}} = \frac{R_O}{(Z_B \| R_C) + R_O} = A_{V0}\left(\frac{1 + s/2\pi z_C}{1 + s/2\pi p_{CX}}\right)$$

$z_C < p_{CX}$ $\quad \angle A_V \leq 90^{\circ}$

$$\text{Phase: } i_B \text{ replenishes } C_O \to \angle A_V = +\tan^{-1}\left(\frac{f_O}{z_C}\right) - \tan^{-1}\left(\frac{f_O}{p_{CX}}\right)$$

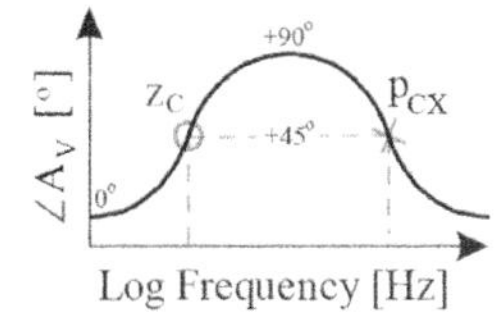

Current-Limit Resistor

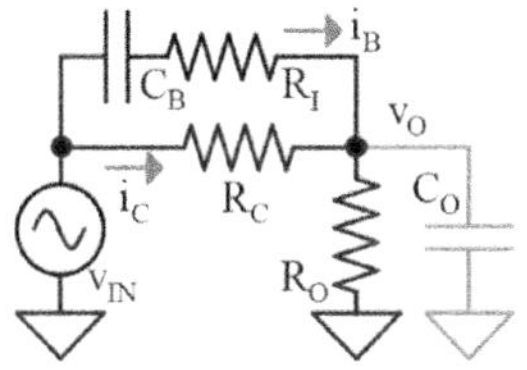

C_B feeds i_B → z_C

R_I limits i_B → Eliminates C_B effects → Reverses z_C

$$A_{V0} = \frac{R_O}{R_C + R_O}$$

C_B & R_I bypass R_C: $\left.\frac{1}{sC_B}\right|_{f_O \geq \frac{1}{2\pi (R_I + R_C) C_B} \equiv z_C} \leq R_I + R_C$

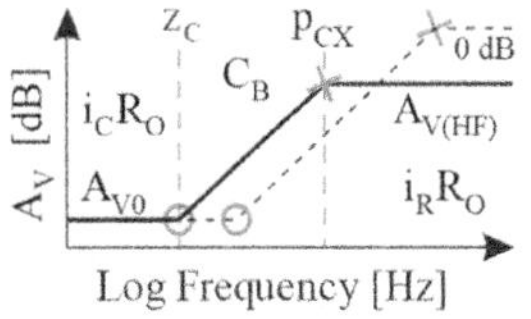

C_B shorts parallel R:

$$\left.\frac{1}{sC_B}\right|_{f_O \geq \frac{1}{2\pi [R_I + (R_C \| R_O)] C_B} \equiv p_{CX}} \leq R_I + (R_C \| R_O) \to A_V\Big|_{f_O > p_{CX}} \approx \frac{i_R R_O}{v_{IN}} = \frac{R_O}{(R_C \| R_I) + R_O}$$

→ Reversal Pole $\qquad$ ($\downarrow A_{V(HF)}$)

$$\text{Gain: } A_V \equiv \frac{v_O}{v_{IN}} = \frac{R_O}{\left[(Z_B + R_I) \| R_C\right] + R_O} = A_{V0}\left(\frac{1 + s/2\pi z_C}{1 + s/2\pi p_{CX}}\right)$$

B. Bypassed Amplifier: In-Phase Capacitor

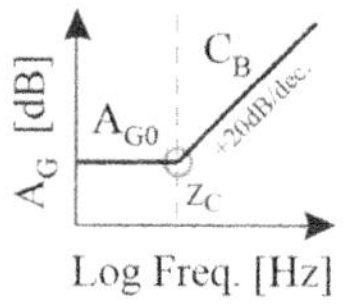

Concept: C_B bypasses amp with f_O → Increasing i_B raises A_G

Embed C_B into Two-Port Model:

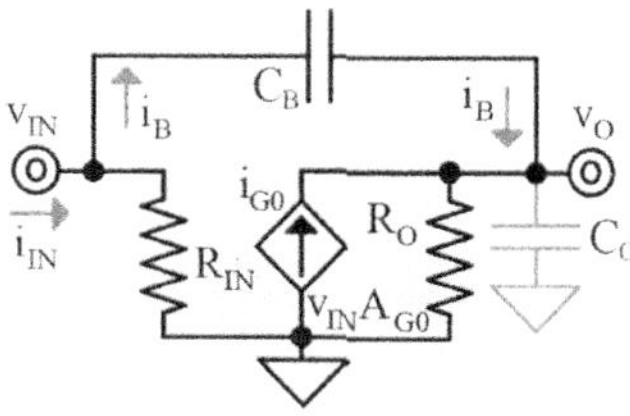

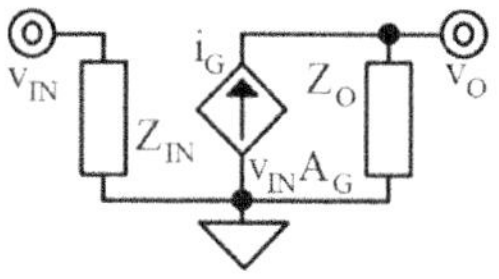

C_B opens at low f_O: $A_G\big|_{Low\ f_O} = A_{G0}$

i_B surpasses i_{G0} past z_C: $i_B = \left.\dfrac{v_{IN} - v_O}{Z_B}\right|_{v_O=0} = v_{IN} sC_B\big|_{f_O \geq \frac{A_{G0}}{2\pi C_B} \equiv z_C} \geq i_{G0} = v_{IN} A_{G0}$

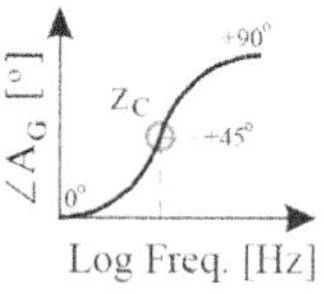

Gain: $A_G \equiv \left.\dfrac{i_G}{v_{IN}}\right|_{v_O=0} \equiv \left.\dfrac{i_{G0} + i_B}{v_{IN}}\right|_{v_O=0} = A_{G0} + \dfrac{1}{Z_B} = A_{G0}\left(1 + \dfrac{sC_B}{A_{G0}}\right) = A_{G0}\left(1 + \dfrac{s}{2\pi z_C}\right)$

i_B replenishes C_O: Recovers up to 90° → $\angle A_G = +\tan^{-1}\left(\dfrac{f_O}{z_C}\right)$

i_B reinforces i_{G0}

In-Phase Zero

Out-of-Phase Capacitor

Inverting Amp: Increasing v_{IN} pulls more i_{G0} → Reduces v_O ∴ i_B opposes i_{G0}

i_B still surpasses i_{G0} past $\dfrac{A_{G0}}{2\pi C_B}$

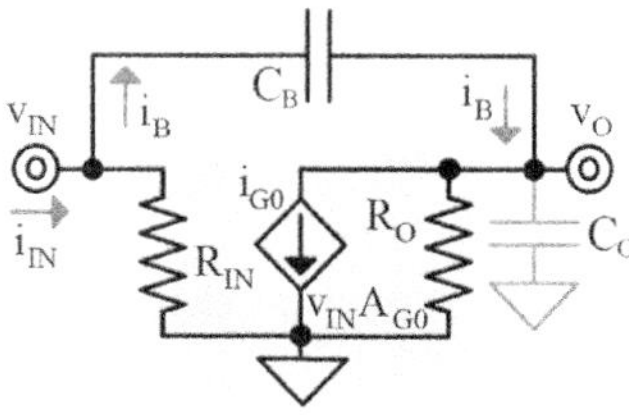

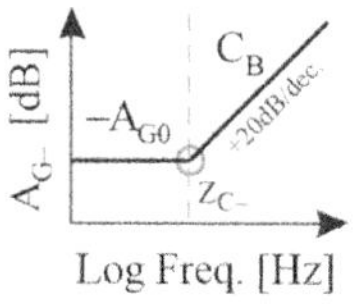

Similar low-f_O Gain = $-A_{G0}$

Except, i_B inverts v_O → Recovers up to 90° & Inverts 180° ∴ Loses up to 90°

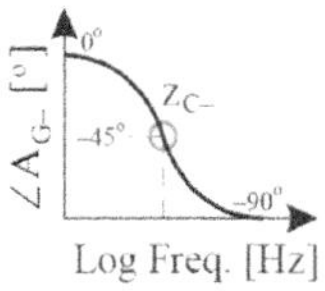

$$\angle A_{G-} = -\tan^{-1}\left(\frac{f_O}{z_{C-}}\right)$$

i_B opposes i_{G0}

Out-of-Phase Zero

Gain: $A_{G-} \equiv \left.\dfrac{i_G}{v_{IN}}\right|_{v_O=0} \equiv \left.\dfrac{i_{G0} - i_B}{v_{IN}}\right|_{v_O=0} = A_{G0} - \dfrac{1}{Z_B} = A_{G0}\left(1 - \dfrac{sC_B}{A_{G0}}\right) = A_{G0}\left(1 - \dfrac{s}{2\pi z_{C-}}\right)$

5.4. Voltage-Sourced LC: A. Series Impedance

Low f_O: L_X shorts & C_O opens $\rightarrow$ $Z_{LC}\big|_{Low\ f_O} \approx Z_C$

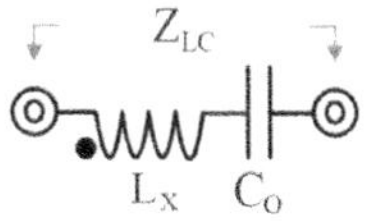

High f_O: L_X opens & C_O shorts $\rightarrow$ $Z_{LC}\big|_{High\ f_O} \approx Z_L$

Z_L overcomes Z_C past f_{LC}: $Z_L = sL_X \geq Z_C = \dfrac{1}{sC_O}$ $\rightarrow$ $f_O \geq \dfrac{1}{2\pi\sqrt{L_X C_O}} \equiv f_{LC}$

At f_{LC}: Z_L & Z_C = Equal complements

$$\rightarrow Z_{LC} = sL_X + \frac{1}{sC_O} = \frac{s^2 L_X C_O + 1}{sC_O}\Bigg|_{f_{LC}} = \frac{-1+1}{(i2\pi f_{LC})C_O} = 0$$

$$s = i2\pi f_{LC} = \frac{i2\pi}{2\pi\sqrt{L_X C_O}}$$

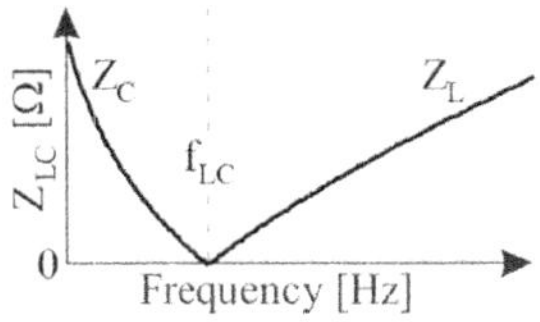

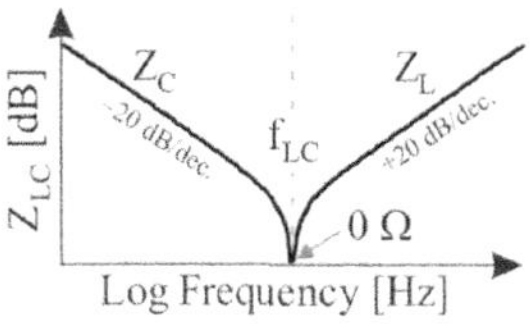

∴ Z_{LC} falls with Z_C — Shorts at f_{LC} — Rises with Z_L past f_{LC}

B. Frequency Response:

i_L rises with $\dfrac{1}{Z_C} \propto f_O$ — Peaks with $\dfrac{1}{R_S}$ at f_{LC} — Falls with $\dfrac{1}{Z_L} \propto \dfrac{1}{f_O}$ past f_{LC}

$$A_G \equiv \frac{i_L}{v_{IN}} = \frac{1}{sL_X + R_S + 1/sC_O} = \frac{sC_O}{s^2 L_X C_O + sR_S C_O + 1} \equiv \frac{1/Z_C}{\left(\frac{s}{2\pi f_{LC}}\right)^2 + \left(\frac{s}{2\pi f_{LC}}\right)\left(\frac{1}{Q_{LC}}\right) + 1}$$

Peak: $A_{G(LC)}\big|_{f_{LC}} = \dfrac{sC_O}{-1 + sR_S C_O + 1}\Bigg|_{f_{LC}} = \dfrac{Q_{LC}}{Z_{C(LC)}} = \dfrac{1}{R_S}$

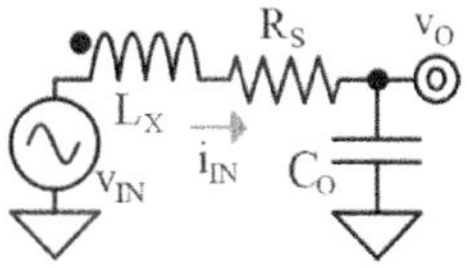

Peaks if C_O shorts above f_{LC} — Q_{LC} over $1/Z_{C(LC)}$

p_{LX} & p_C $\rightarrow$ p_{LC}

LC Quality: $Q_{LC} \equiv \dfrac{f_{CS}}{f_{LC}} = \dfrac{f_{LC}}{f_{LS}} = \dfrac{1}{R_S}\sqrt{\dfrac{L_X}{C_O}} > 1$

If L_X overcomes R_S below f_{LC}

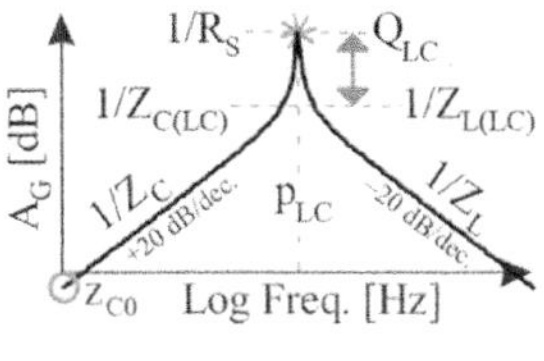

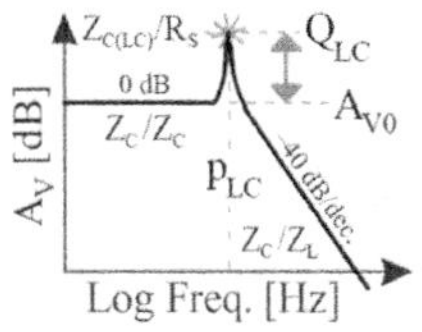

∴ Peaks when L_X & C_O interact at f_{LC} $\rightarrow$ $Z_{L(LC)} = Z_{C(LC)} > R_S$

$A_V \equiv \dfrac{v_O}{v_{IN}} = A_G Z_C$

C. Resistive Effects: Peaked

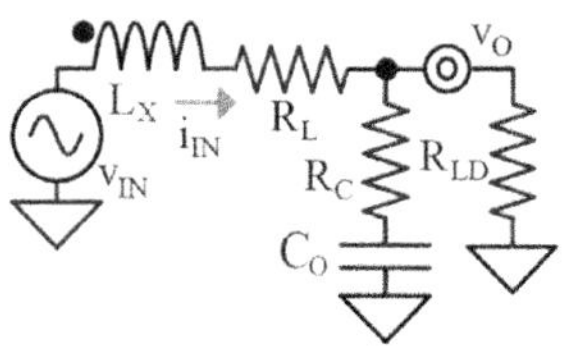

Load R_{LD} & parasitic series R_L & R_C

R_L reduces low-f_O gain: $A_{V0} = A_{G0}R_{LD} \approx \dfrac{R_{LD}}{R_L + R_{LD}}$

L_X & C_O interact when Z_L & Z_C dominate (eclipse R_L, R_C, R_{LD}):

When Z_L overcomes R_L below f_{LC}: $sL_X\Big|_{f_O \geq \frac{R_L}{2\pi L_X} \equiv f_L} \geq R_L$

When C_O & R_C shunt R_{LD} below f_{LC}: $\dfrac{1}{sC_O}\Bigg|_{f_O \geq \frac{1}{2\pi(R_C + R_{LD})C_O} \equiv f_{CP}} \leq R_C + R_{LD}$

When C_O shorts above f_{LC}: $\dfrac{1}{sC_O}\Bigg|_{f_O \geq \frac{1}{2\pi R_C C_O} \equiv f_C} \leq R_C$

Example: Determine A_{V0}, $A_{V(LC)}$, poles, & zeros when $L_X = 10\ \mu H$, $R_L = 50\ m\Omega$, $C_O = 5\ \mu F$, $R_C = 10\ m\Omega$, $R_{LD} = 100\ \Omega$.

Solution:

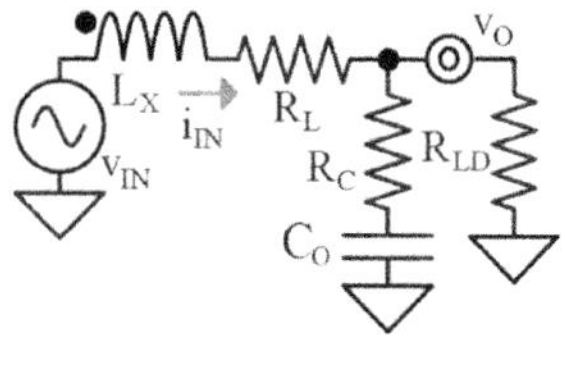

$$f_L = \frac{R_L}{2\pi L_X} = \frac{50m}{2\pi(10\mu)} = 800\ \text{Hz}$$

$$f_{CP} = \frac{1}{2\pi\left(R_C + R_{LD}\right)C_O} = \frac{1}{2\pi(10m + 100)(5\mu)} = 320\ \text{Hz}$$

$$f_{LC} = \frac{1}{2\pi\sqrt{L_X C_O}} = \frac{1}{2\pi\sqrt{(10\mu)(5\mu)}} = 22\ \text{kHz}$$

$$A_{V0} \approx \frac{R_{LD}}{R_L + R_{LD}} = 1.0\ \text{V/V}$$

$$f_C = \frac{1}{2\pi R_C C_O} = \frac{1}{2\pi(10m)(5\mu)} = 3.2\ \text{MHz}$$

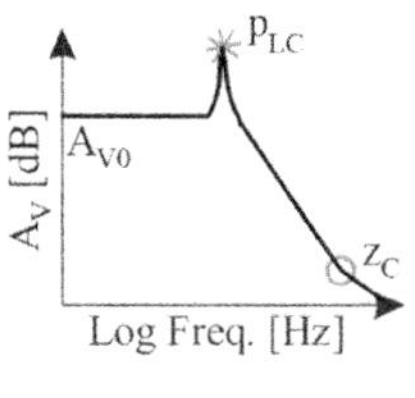

$$f_{CP} < f_L < f_{LC} < f_C \quad \therefore \quad p_L = p_C = p_{LC} = f_{LC} < z_C = f_C$$

$$A_{V(LC)} \approx \frac{Z_{CO(LC)} \parallel R_{LD}}{R_L + \left(R_C \parallel R_{LD}\right)} = \left[\frac{1}{50m + (10m \parallel 100)}\right]\left(\frac{1}{2\pi(22k)(5\mu)} \parallel 100\right) = 24\ \text{V/V}$$

Unpeaked Scenarios: $Q_{LC} \le 1$ → $R_L > Z_{L/C(LC)}$

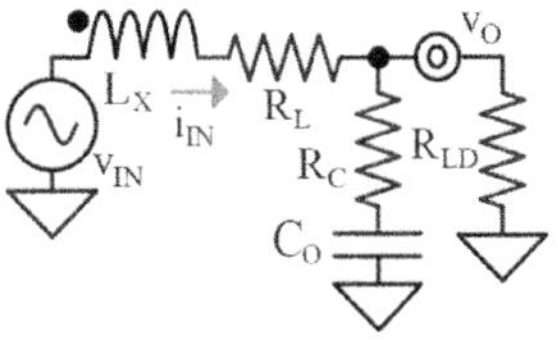

i. If L_X cannot overcome R_L below f_{LC} → $L_X \approx$ Short at f_{LC}

$$C_O \text{ \& } R_C \text{ shunt } R_{LD} \,||\, R_L \text{ past } p_C \approx f_{CS} \equiv \frac{1}{2\pi\left[R_C + \left(R_{LD} \,||\, R_L\right)\right]C_O} < f_{LC}$$

↳ Shorts above f_{LC}

$$L_X \text{ overcomes } R_L + (R_C \,||\, R_{LD}) \text{ past } p_L \approx f_{LS} \equiv \frac{R_L + \left(R_C \,||\, R_{LD}\right)}{2\pi L_X} > f_{LC}$$

ii. If C_O & R_C cannot shunt R_{LD} below f_{LC} → $R_{LD} < Z_{L/C(LC)}$ $C_O \approx$ Open at f_{LC}

$$L_X \text{ overcomes } R_L + R_{LD} \text{ past } p_L \approx f_{LP} \equiv \frac{R_L + R_{LD}}{2\pi L_X} < f_{LC}$$

↳ Opens above f_{LC}

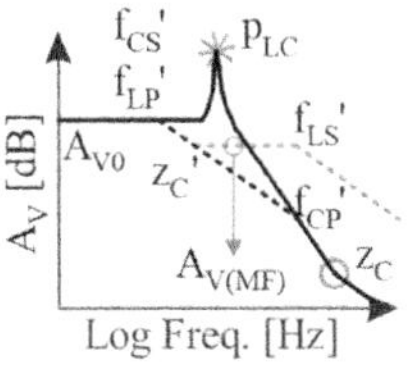

$$C_O \text{ \& } R_C \text{ shunt } R_{LD} \text{ past } p_C \approx f_{CP} \equiv \frac{1}{2\pi\left(R_C + R_{LD}\right)C_O} > f_{LC}$$

iii. If C_O shunts & shorts below f_{LC} → L_X overcomes $R_L + (R_C \,||\, R_{LD})$ past $p_L \approx f_{LS}$

Example: Determine A_{V0}, $A_{V(LC)}$, poles, & zeros when $L_X = 10\ \mu H$, $R_L = 10\ \Omega$, $C_O = 5\ \mu F$, $R_C = 10\ m\Omega$, $R_{LD} = 100\ \Omega$.

Solution: $f_{LC} = 22$ kHz, $f_C = 3.2$ MHz from previous example

$$A_{V0} = \frac{R_{LD}}{R_L + R_{LD}} = \frac{100}{10+100} = 910\ \text{mV/V}$$

$$f_L = \frac{R_L}{2\pi L_X} = \frac{10}{2\pi(10\mu)} = 160\ \text{kHz} > f_{LC}$$

↓ $L_X \approx$ Short below f_{LC}

$$f_{CS} = \frac{1}{2\pi\left[R_C + \left(R_{LD} \,||\, R_L\right)\right]C_O} = \frac{1}{2\pi\left[10m + (100 \,||\, 10)\right](5\mu)} = 3.5\ \text{kHz}$$

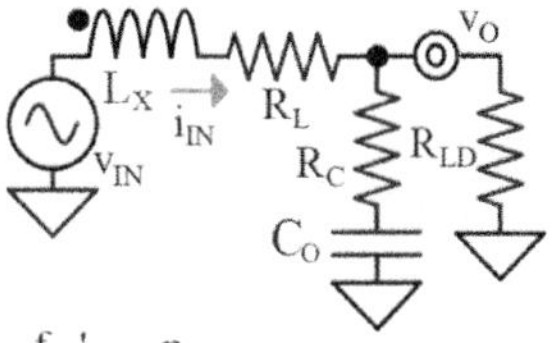

$$f_{LS} = \frac{R_L + \left(R_C \,||\, R_{LD}\right)}{2\pi L_X} = \frac{10 + \left(10m \,||\, 100\right)}{2\pi(10\mu)} = 160\ \text{kHz}$$

∴ $f_{CS} \approx p_C < f_{LC} < f_{LS} \approx p_L < f_C = z_C$ → No peak

↳ C_O shunts ↳ L_X impedes ↘ C_O shorts

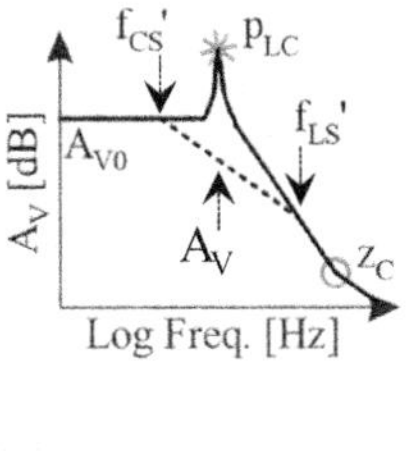

$$A_{V(LC)} \approx A_{V0}\left(\frac{f_{CS}}{f_{LC}}\right) = (910m)\left(\frac{3.5k}{22k}\right) = 140\ \text{mV/V}$$

D. Phase Shift

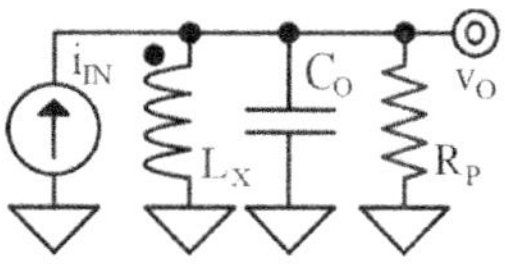

v_O or i_L: Rises with Z_L or $\frac{1}{Z_C}$ Falls with Z_C or $\frac{1}{Z_L}$ past f_{LC}

Peaks when L_X & C_X swap dominance at f_{LC}

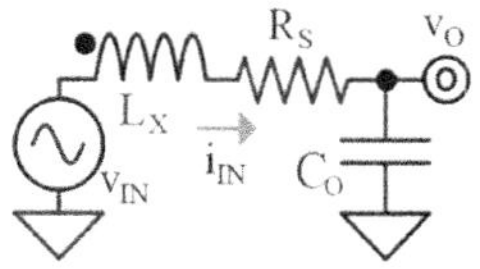

At f_{LC}: Phase loses 90° that sL_X or sC_O in $z_{L/C0}$ added

90° that $1/sC_O$ or $1/sL_X$ in $p_{C/L}$ deducts

} p_{LC} deducts 180°

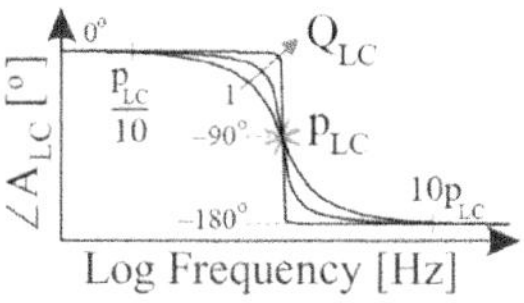

Phase Loss: Abrupt with high Q_{LC}

Gradual with low Q_{LC} → Resistances dampen transition

$$\text{Tan}^{-1} = \pm 90^\circ \text{ with } \pm \text{ operands} \quad \therefore \quad \angle A_{LC} = \tan^{-1}\left[Q_{LC}\left(\frac{f_{LC}}{f_O} - \frac{f_O}{f_{LC}}\right)\right] - 90^\circ$$

Q_{LC}: Amplifies difference; $\left(\frac{f_{LC}}{f_O} - \frac{f_O}{f_{LC}}\right)$: Determines polarity; -90°: Fixes Offset

SPICE Simulation

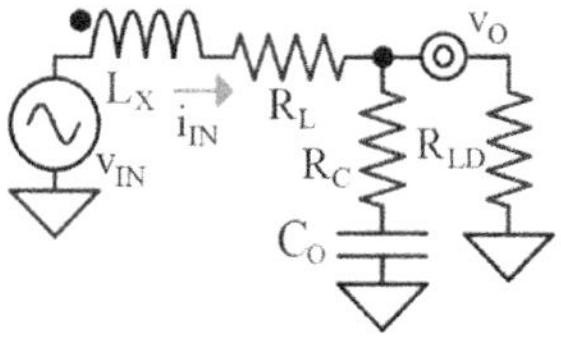


```
* Voltage-Sourced LC
vin vin 0 dc=0 ac=1
lx vin vl 10u
rl vl vo 50m
*rl vl vo 10
*rl vl vo 50
rc vo vc 10m
*rc vo vc 10
co vc 0 5u
rld vo 0 100
*rld vo 0 1.4
.ac dec 1000 10 100e6
.end
```

(ac=1) → Small sinusoid at every f_O

(.ac dec 1000 10 100e6) → 1000 pts / decade from 10 Hz to 100 MHz

* Plot v_O in dB → $A_V = v_O / v_{IN} = v_O / 1 = v_O$

* High R_L → $p_C \approx f_{CS} < f_{LC} < p_L \approx f_{LS}$

* High R_C → $p_C \le z_C < f_{LC} < p_L$ No peak

* $R_{LD} = 1.4\ \Omega$ → $f_{CS} = f_{LC}$ ∴ $Q_{LC} = 1$

5.5. Switched Inductor: A. Signal Translations

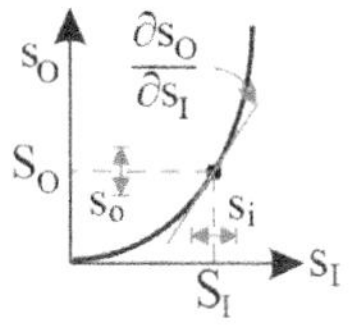

Aim: Keep v_O or i_O near target → Small variations

How: Feedback loop senses & uses small variations to adjust d_E

Signal: Small dynamic component ≡ Lower-case variables & subscripts: s_o

Static steady-state component ≡ Upper-case variables & subscripts: S_O

Complete ≡ Lower-case variables & upper-case subscripts: s_O

Analysis: Linear projections approximate small-signal translations

Switched Inductors: v_L pulses volts & i_L ripples amps

$$s_o = s_i A_X \approx s_i \underbrace{\left(\frac{\partial s_O}{\partial s_I}\right)}_{\text{Slope}}$$

Static Periodic State: Cycles repeat (variations fade)

Feedback Signals: Small variations in amplitude, d_E, f_{SW}

B. Small-Signal Model

Operation: Switches energize & drain L_X with v_{IN}, v_O, ground.

d_E' commands switches to energize t_E fraction of t_{SW} period.

Two-Port Model:

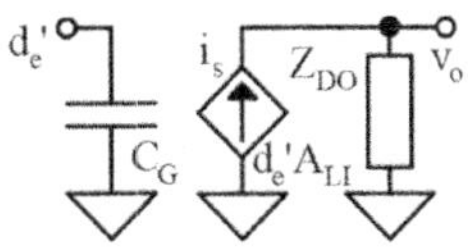

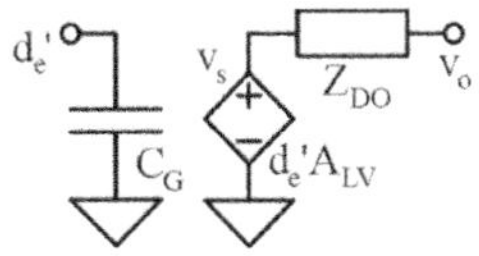

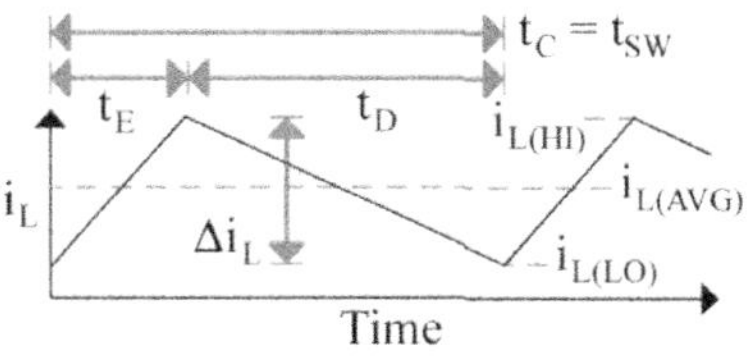

Continuous Conduction:

$d_E' = d_E = t_E$ of t_{SW}

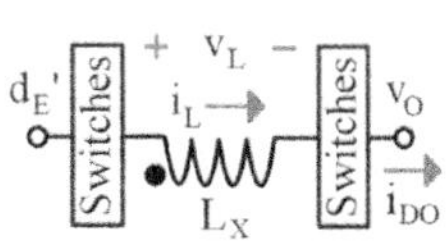

i_L rises when v_E energizes L_X & falls when v_D drains L_X.

S_{DO} duty-cycles i_L into v_O → $d_O = 1$ in bucks $d_O = d_D$ with S_{DO}

Output Impedance

Test Condition: Disable A_{LI}, A_{LV} → No d_E' variations → $d_e' = 0$

$$\therefore \quad t_E, t_D, t_O, t_{SW} = T_E, T_D, T_O, T_{SW} = \text{Static}$$

Duty-Cycled v_O: L_X & R_L load v_o across T_O what L_{DO} & R_{LO} load across T_{SW}

$$\rightarrow \quad E_{LX} = \left(\frac{1}{2}\right) L_X i_{LX}^{\ 2} = \left(\frac{1}{2}\right) L_X \left[\left(\frac{v_o}{L_X}\right) T_O\right]^2 = \left(\frac{v_o^{\ 2}}{2}\right)\left(\frac{D_O^{\ 2}}{L_X}\right) T_{SW}^{\ 2}$$

$$\equiv E_{LO} = \left(\frac{1}{2}\right) L_{DO} i_{LO}^{\ 2} = \left(\frac{1}{2}\right) L_{DO} \left[\left(\frac{v_o}{L_{DO}}\right) T_{SW}\right]^2 = \left(\frac{v_o^{\ 2}}{2}\right)\left(\frac{1}{L_{DO}}\right) T_{SW}^{\ 2}$$

$$\therefore \quad L_{DO} = \frac{L_X}{D_O^{\ 2}}$$

$$\rightarrow \quad P_{RL} = \left(\frac{v_o^{\ 2}}{R_L}\right)\left(\frac{T_O}{T_{SW}}\right) = \frac{v_o^{\ 2}}{R_L / D_O} \equiv P_{RO} = \frac{v_o^{\ 2}}{R_{LO}} \qquad \therefore \quad R_{LO} = \frac{R_L}{D_O}$$

Gain: Inductor Current

Test Condition: Disable Z_{DO} → No v_O variations → $v_o = 0$

Independent Source: No v_{IN} variations → $v_{in} = 0$

(together: $v_e = 0$, $v_d = 0$)

Ohmic Translation: $$i_L = \frac{v_L}{Z_L} = \frac{v_E d_E - v_D d_D}{Z_L} = \frac{v_E d_E - v_D(1-d_E)}{Z_L} = \frac{(v_E + v_D) d_E - v_D}{Z_L}$$

Static Variation: $$d_E = \frac{v_D}{v_E + v_D} \quad \therefore \quad v_L = i_L Z_L = 0$$ → L_X shorts at low f_O

Dynamic Variation: $$i_l' \equiv i_l\big|_{v_o=0} = d_e' A_L = d_e' \left(\frac{\partial i_L}{\partial d_E}\right)\bigg|_{v_o=0} = d_e' \left(\frac{V_E + V_D}{Z_L}\right)$$

q_d delivers $i_d = i_l'$ averaged across T_{SW} & delivered across $T_O \approx i_l'D_O$

$$i_{do} \text{ loses } i_e = \frac{q_e}{T_{SW}} \approx \frac{t_e I_{L(HI)}}{T_{SW}} = d_e' I_{L(HI)}$$

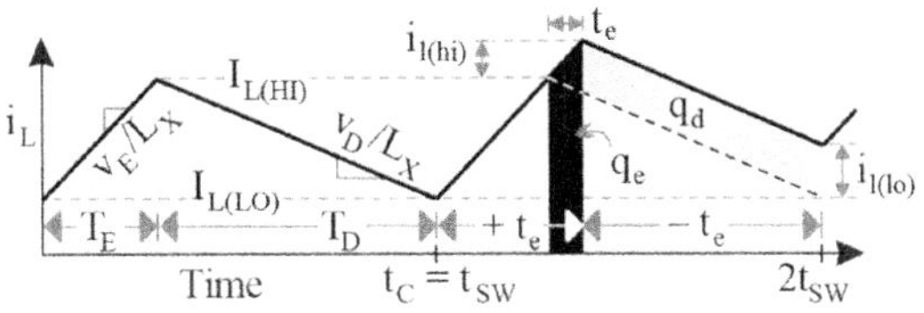

i_d falls with $\frac{1}{Z_L}$ $\quad\therefore\quad$ i_e overcomes i_d & inverts i_{do} past out-of-phase zero z_{DO}:

$$i_d \approx i_l'D_O = d_e'\left(\frac{V_E+V_D}{sL_X}\right)D_O\Bigg|_{f_O \ge \left(\frac{V_E+V_D}{2\pi L_X}\right)\left(\frac{D_O}{I_{L(HI)}}\right) \equiv z_{DO} \approx \left(\frac{V_E+V_D}{2\pi L_X}\right)\left(\frac{D_O^2}{I_O}\right) \equiv \frac{R_{LD}'}{2\pi L_X}} \le i_e \approx d_e' I_{L(HI)}$$

Signal-Flow Graph

$$\therefore \quad A_{LI} \equiv \frac{i_s}{d_e'}\bigg|_{v_o=0} \approx \left(\frac{V_E+V_D}{Z_L}\right)D_O\left(1-\frac{s}{2\pi z_{DO}}\right)$$

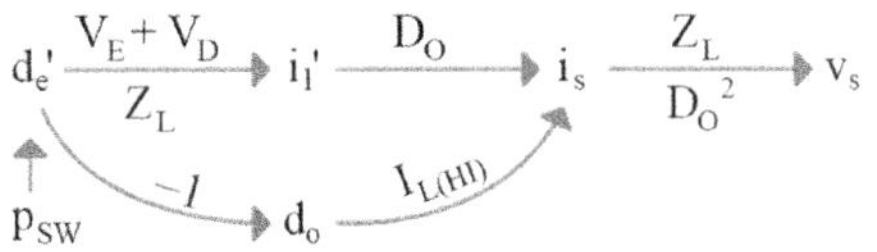

$$A_{LV} \equiv \frac{v_s}{d_e'}\bigg|_{i_o=0} = A_{LI}Z_{LO} = A_{LI}\left(\frac{Z_L}{D_O^2}\right) \approx \left(\frac{V_E+V_D}{D_O}\right)\left(1-\frac{s}{2\pi z_{DO}}\right) \ne f(Z_L)$$

→ Better model

Discontinuous Conduction

$$d_E' = \frac{t_E}{t_{SW}} \ne d_E = \frac{t_E}{t_C}$$

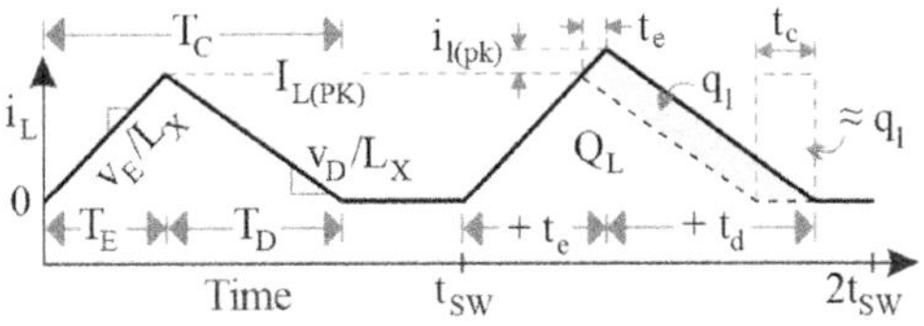

L_X energizes & depletes every cycle $\quad\therefore\quad$ i_s delivers all of i_l' → No z_{DO} loss

t_D scales with t_E

Static v_E & v_D projections

} d_e & d_d scale proportionally } $\frac{v_l}{Z_L} = \frac{V_E d_e - V_D d_d}{Z_L} = 0$

$\therefore \quad i_{do} \ne f(f_O)$ → No inductive effect → $Z_{DO} = R_{DO}$

t_e & t_c scale with T_E & T_C → $d_E = \frac{T_E + t_e}{T_C + t_c} = \frac{T_E}{T_C} = \frac{t_e}{t_c} = D_E$ → $d_e = 0$

t_e projects proportional $i_{l(pk)}$ → Small-signal fraction $k_d \equiv \frac{t_e}{T_E} = \frac{t_c}{T_C} = \frac{i_{l(pk)}}{I_{L(PK)}}$

Output Impedance

Test Condition: Disable A_{LI}, A_{LV} → No d_E' variations → $d_e' = 0$

$$\therefore \quad t_E, t_D, t_O, t_{SW} = T_E, T_D, T_O, T_{SW} = \text{Static}$$

Duty-Cycled Output: L_X loads v_o across T_O what R_{DO} loads across T_{SW}

$$\rightarrow \quad E_{LX} = \left(\frac{1}{2}\right)L_X i_{LX}^2 = \left(\frac{1}{2}\right)L_X\left[\left(\frac{v_o}{L_X}\right)D_O T_C\right]^2 = \left(\frac{v_o^2}{2}\right)\left(\frac{D_O^2}{L_X}\right)T_C^2$$

$$P_{RO} = \frac{v_o^2}{R_{DO}} \equiv \frac{E_{LX}}{T_{SW}} = \left(\frac{v_o^2}{2}\right)\left(\frac{D_O^2}{L_X}\right)\left(\frac{T_C^2}{T_{SW}}\right)$$

R_L raises R_{DO}, but usually not by much:

$$\therefore \quad R_{DO} = 2\left(\frac{L_X}{D_O^2}\right)\left(\frac{T_{SW}}{T_C^2}\right) + \frac{R_L}{D_O} = 2L_{DO}\left(\frac{T_{SW}}{T_C^2}\right) + R_{LO} \neq f(f_O)$$

Gain: Inductor Current

i_s delivers $q_l = q_L - Q_L$ → Difference of triangle areas

$$= 0.5t_C i_{L(PK)} - 0.5T_C I_{L(PK)} = 0.5T_C(1 + k_d)I_{L(PK)}(1 + k_d) - 0.5T_C I_{L(PK)}$$

$$= 0.5T_C I_{L(PK)}(k_d^2 + 2k_d) \approx T_C I_{L(PK)} k_d = t_c I_{L(PK)} \quad \leftarrow \quad k_d = \text{Small fraction}$$

$$i_s = i_{do}\Big|_{v_o=0} = \frac{q_l}{T_{SW}} \approx \frac{t_c I_{L(PK)}}{T_{SW}} = \frac{t_e I_{L(PK)}}{D_E T_{SW}} = d_e'\left(\frac{I_{L(PK)}}{D_E}\right)$$

$$A_{LI} \equiv \frac{i_s}{d_e'}\Big|_{v_o=0} \approx \frac{I_{L(PK)}}{D_E} \neq f(L_X)$$

→ Good model

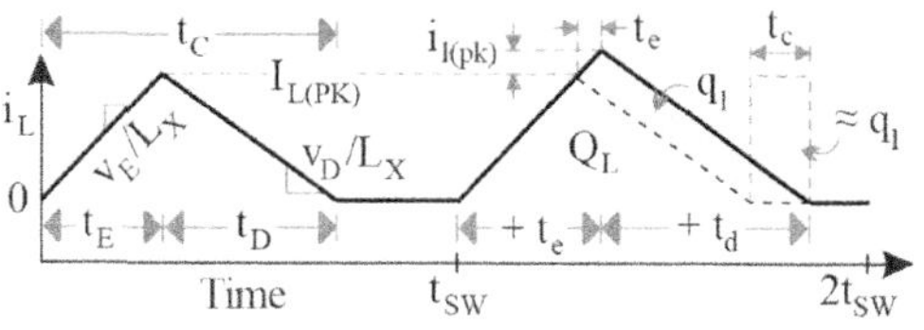

Switching Pole:

Switcher waits up to t_{SW} to adjust next t_E

One t_E adjustment per cycle

} Delays & masks sub-cycle (higher-f_O) variations → Model with $p_{SW} \approx f_{SW}$

C. Power Stage: Continuous Conduction

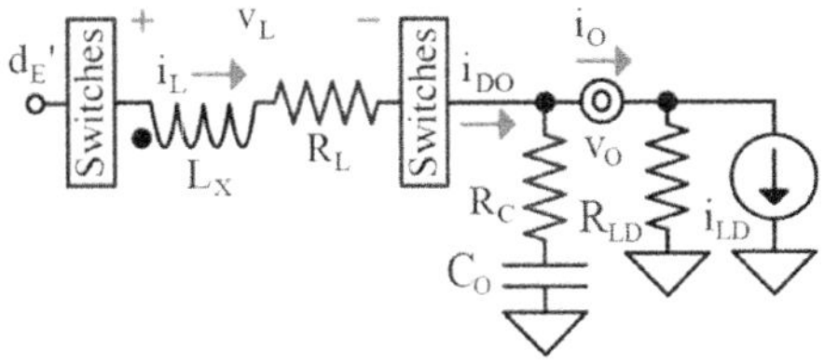

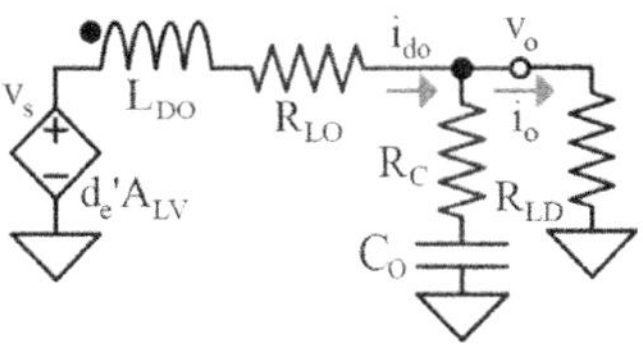

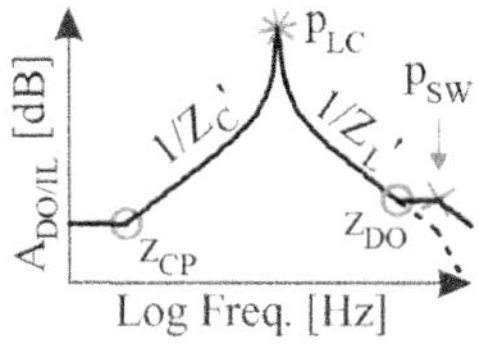

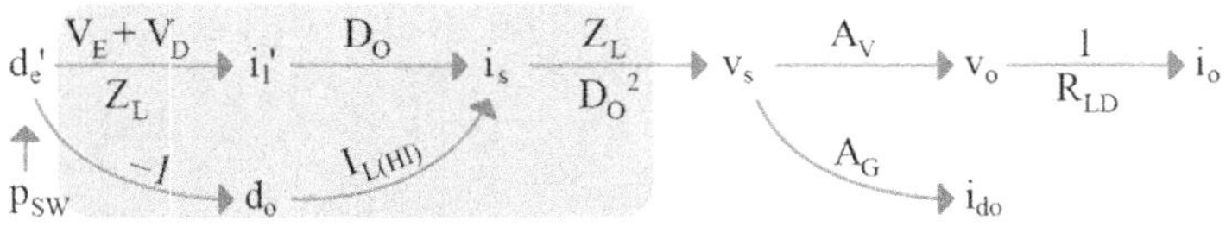

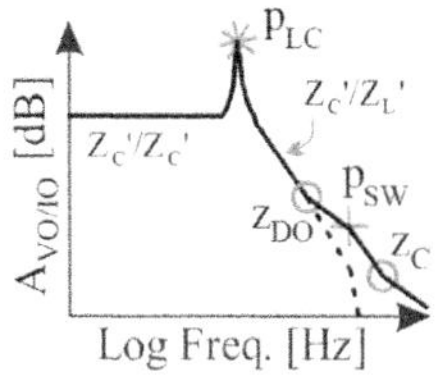

Gains:

$$A_{DO} \equiv \frac{i_{do}}{d_e'} = A_{LV}A_G = \left(\frac{V_E + V_D}{D_O}\right)\left(1 - \frac{s}{2\pi z_{DO}}\right)A_G$$

$$A_{VO} \equiv \frac{v_o}{d_e'} = A_{DO}\left[\left(Z_C + R_C\right) \| R_{LD}\right]$$

Reverse D_O translation of i_{do} without z_{DO}

$$A_{IL} \equiv \frac{i_l}{d_e'} = \left(\frac{V_E + V_D}{D_O}\right)\left(\frac{A_G}{D_O}\right)$$

$$A_{IO} \equiv \frac{i_o}{d_e'} = \frac{A_{VO}}{R_{LD}}$$

Example: Determine A_{VO0}, poles, & zeros in CCM when $v_E = 2$ V, $v_D = 4$ V, $d_E = 67\%$, $d_O = 33\%$, $i_{L(HI)} = 190$ mA, $t_{SW} = 1$ µs, $L_X = 10$ µH, $R_L = 50$ mΩ, $C_O = 5$ µF, $R_C = 10$ mΩ, $R_{LD} = 100$ Ω.

Solution: $f_{CP} = 320$ Hz, $f_C = 3.2$ MHz from earlier example

$$A_{VO0} = A_{LV0}A_{V0} \approx \left(\frac{V_E + V_D}{D_O}\right)\left[\frac{R_{LD}}{\left(R_L/D_O\right) + R_{LD}}\right] = 18 \text{ V/V}$$

$$f_L = \frac{R_L/D_O}{2\pi\left(L_X/D_O^2\right)} = \frac{50m/33\%}{2\pi(10\mu/33\%^2)} = 260 \text{ Hz}$$

$$f_{SW} = \frac{1}{t_{SW}} = \frac{1}{1\mu} = 1 \text{ MHz}$$

$$f_{LC} = \frac{1}{2\pi\sqrt{\left(L_X/D_O^2\right)C_O}} = \frac{1}{2\pi\sqrt{(10\mu/33\%^2)(5\mu)}} = 7.4 \text{ kHz}$$

$\therefore \quad f_L \approx f_{CP} < p_{LC} = f_{LC} < z_{DO} < p_{SW} \approx f_{SW} < z_C = f_C$

No R_{LO} ↵ ↳ No R_{LD} No R_C below f_{LC} ↵

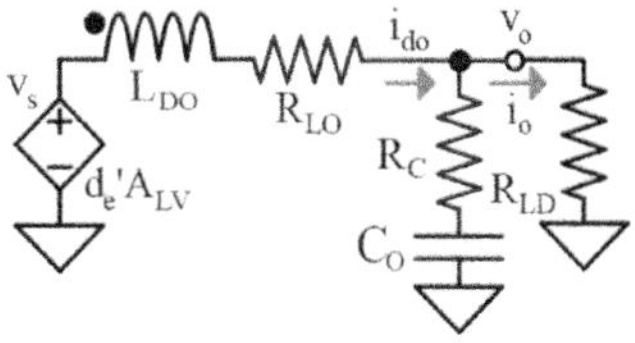

$$z_{DO} = \left(\frac{V_E + V_D}{2\pi L_X}\right)\left(\frac{D_O}{I_{L(HI)}}\right) = \left[\frac{2+4}{2\pi(10\mu)}\right]\left(\frac{33\%}{190m}\right) = 170 \text{ kHz}$$

C. Power Stage: Discontinuous Conduction

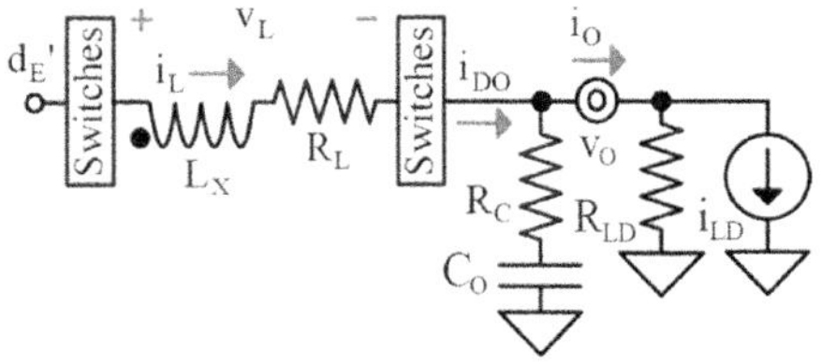

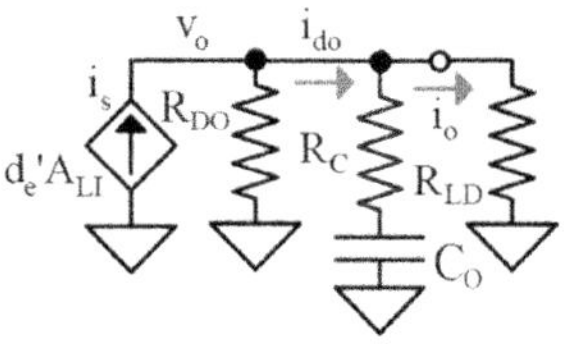

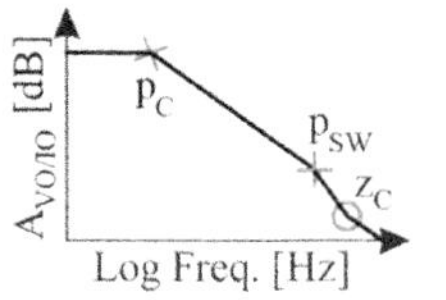

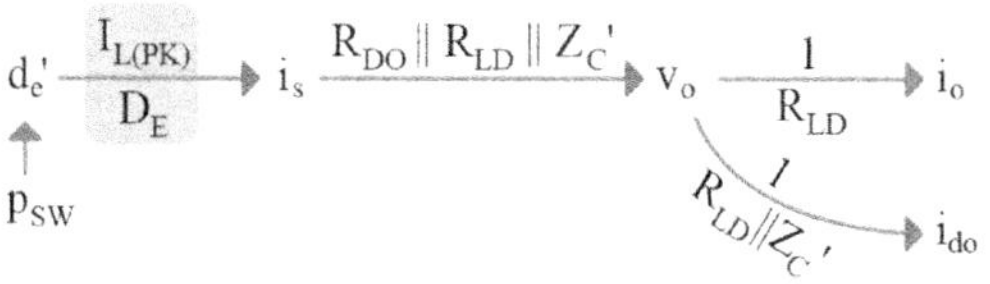

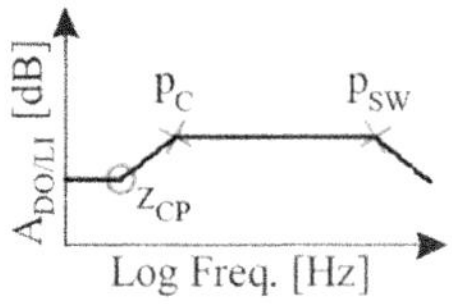

Gains:

$$A_{VO} \equiv \frac{v_o}{d_e'} = A_{LI}\left[R_{DO} \parallel R_{LD} \parallel \left(Z_C + R_C\right)\right] \approx \left(\frac{I_{L(PK)}}{D_E}\right)\left\{\frac{\left(R_{DO} \parallel R_{LD}\right)\left(1+sR_C C_O\right)}{1+s\left[R_C+\left(R_{DO} \parallel R_{LD}\right)\right]C_O}\right\}$$

$$A_{IO} \equiv \frac{i_o}{d_e'} = \frac{A_{VO}}{R_{LD}} \qquad A_{DO} \equiv \frac{i_{do/l}}{d_e'} = \frac{A_{VO}}{R_{LD} \parallel \left(Z_C + R_C\right)} = \left(\frac{A_{VO}}{R_{LD}}\right)\left[\frac{1+s\left(R_C+R_{LD}\right)C_O}{1+sR_C C_O}\right]$$

Example: Determine A_{VO0}, poles, & zeros in DCM when $d_E = d_O = 50\%$,

$t_C = 570$ ns, $t_{SW} = 1$ µs, $i_{L(PK)} = 57$ mA, $L_X = 10$ µH, $R_L = 50$ mΩ,

$C_O = 5$ µF, $R_C = 10$ mΩ, $R_{LD} = 500$ Ω.

Solution: $p_{SW} \approx f_{SW} = 1$ MHz, $z_C = f_C = 3.2$ MHz from earlier examples

$$R_{DO} = 2\left(\frac{L_X}{D_O^{\,2}}\right)\left(\frac{T_{SW}}{T_C^{\,2}}\right) + \frac{R_L}{D_O} = 2\left(\frac{10\mu}{50\%^2}\right)\left(\frac{1\mu}{570n^2}\right) + \frac{50m}{50\%} = 250\ \Omega$$

$$A_{VO0} = \left(\frac{I_{L(PK)}}{D_E}\right)\left(R_{DO} \parallel R_{LD}\right) \approx \left(\frac{57m}{50\%}\right)\left(250 \parallel 500\right) = 18\ \text{V/V}$$

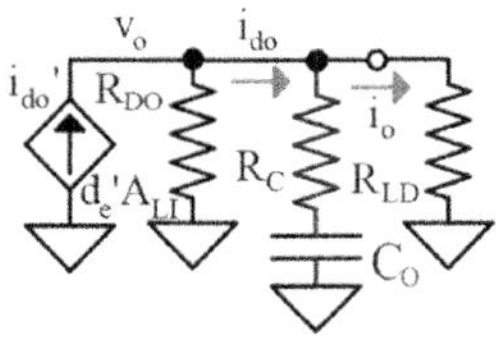

$$p_C = \frac{1}{2\pi\left[R_C+\left(R_{DO} \parallel R_{LD}\right)\right]C_O} = \frac{1}{2\pi[10m+(250 \parallel 500)](5\mu)} = 190\ \text{Hz}$$

Chapter 6. Feedback Control

6.1. Negative Feedback

6.2. Op-Amp Translations

6.3. Stabilizers

6.4. Voltage Control

6.5. Current Control

6.6. Digital Control

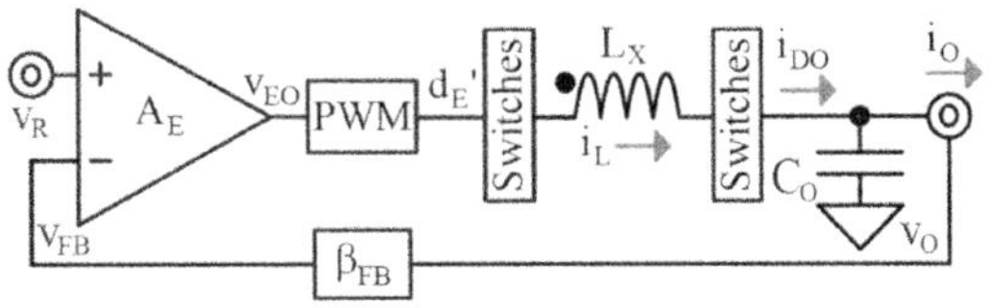

6.1. Negative Feedback: A. Model

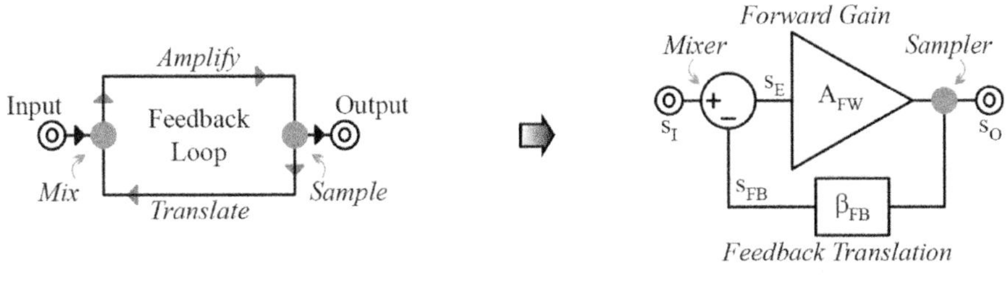

Mixer: s_I s_{FB} s_E → Same type ∴ Same dimensional units

Gains: $A_{FW} \equiv \dfrac{s_O}{s_E}$ $\beta_{FB} \equiv \dfrac{s_{FB}}{s_O}$ Loop Gain $\equiv A_{LG} = \dfrac{s_{FB}}{s_E} = A_{FW}\beta_{FB}$

B. Closed-Loop Translations

Error: $s_E = s_I - s_{FB} = s_I - s_E A_{FW}\beta_{FB} = s_I - s_E A_{LG} = \dfrac{s_I}{1 + A_{LG}}$

If $A_{LG} \rightarrow \infty \quad \therefore \quad s_E \rightarrow 0$ → A_{LG} suppresses s_E

Gain error $s_E \approx \dfrac{s_I}{A_{LG}}$

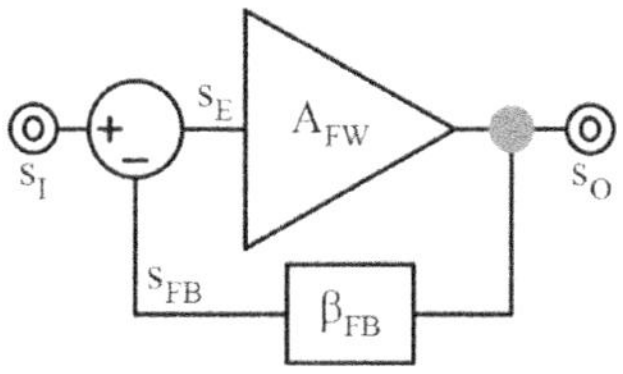

$s_{FB} \approx s_I$ → Mirrored reflection

– FB opposes deviations

$s_O = \dfrac{s_{FB}}{\beta_{FB}} \approx \dfrac{s_I}{\beta_{FB}}$ → Fwd gain translation $\approx \dfrac{1}{\beta_{FB}}$

Gain: $s_O = s_E A_{FW} = \left(s_I - s_{FB}\right)A_{FW} = \left(s_I - s_O\beta_{FB}\right)A_{FW} = \dfrac{s_I A_{FW}}{1 + A_{FW}\beta_{FB}} = \dfrac{s_I A_{FW}}{1 + A_{LG}}$

$$A_{CL} \equiv \frac{s_O}{s_I} = \frac{A_{FW}}{1 + A_{FW}\beta_{FB}} = A_{FW} \,||\, \frac{1}{\beta_{FB}} \approx \text{Lowest Forward Translation}$$

C. Frequency Response

Response: A_{CL} follows lowest forward translation

Conversion: Poles & Zeros in β_{FB} = Zeros & Poles in $\dfrac{1}{\beta_{FB}}$ → Opposite effect

Inclusion: Poles & zeros in lowest forward translation appear in A_{CL}

Crossings: Poles or zeros appear or disappear → E.g.'s: p_{X1}, p_{X2}, p_{X34}

Projections: Each pole/zero lowers/raises gain 20 dB/dec. → E.g. 1: $\dfrac{p_{X1}}{z_1} = \dfrac{1/\beta_{FB}}{A_{FW0}}$

E.g. 1:

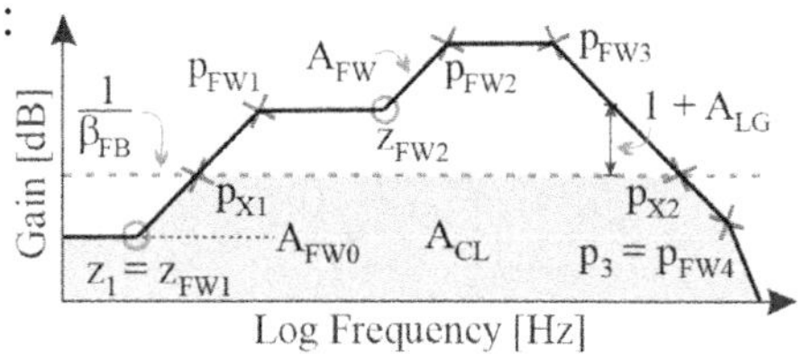

E.g. 2:

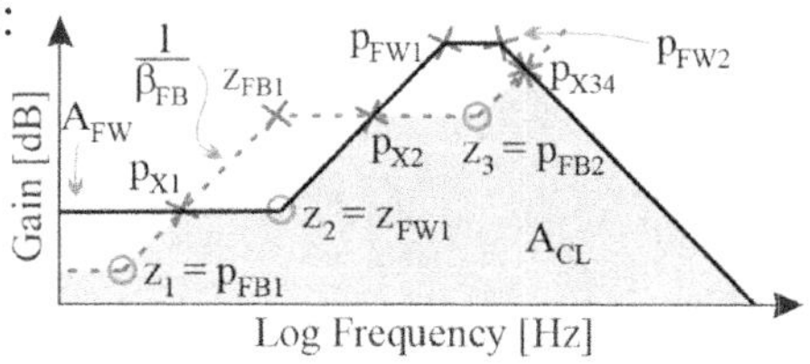

D. Stability

Gain Objective: $A_{CL} \approx \frac{1}{\beta_{FB}} \geq 1 \quad \therefore \quad A_{FW} > \frac{1}{\beta_{FB}} \quad \beta_{FB} \leq 1$

Stability Criterion: If $p_1 \; p_2 < f_{180°} \equiv$ Inversion $f_O < f_{0dB} \equiv$ Unity-Gain f_O

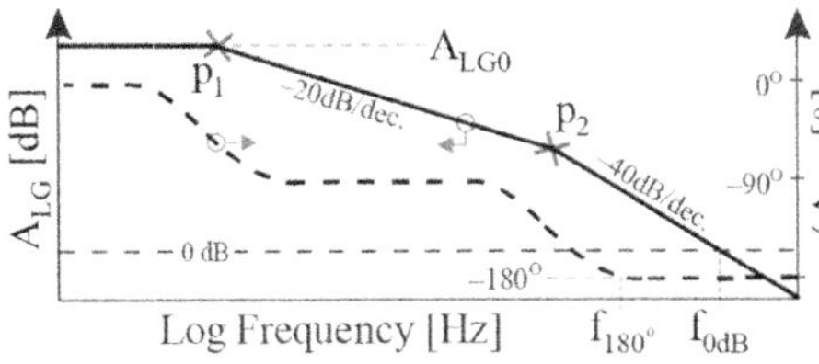

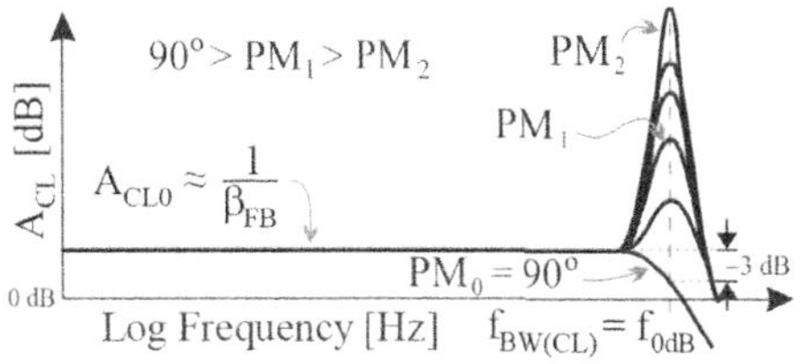

$A_{LG(0dB)} = 1\angle{-180°}$

$\downarrow$

$\therefore \quad A_{CL} = A_{FW} \,||\, \frac{1}{\beta_{FB}} = \frac{A_{FW}}{1 + A_{LG}} = \frac{A_{FW}}{1 - 1} \to \infty$ → Uncontrolled

Loop inverts at $f_{180°}$ → + FB

"Stable" if phase shift < 180°

$f_{BW(CL)} = f_{0dB}$

$f_{0dB} < f_{180°}$

$\downarrow$

Phase Margin $\equiv 180° + \angle A_{LG(0dB)} > 0°$

Gain Margin $\equiv 0 \text{ dB} - A_{LG(180°)} > 0 \text{ dB}$

Stabilization

Objective: Reach f_{0dB} with < 180°

Below f_{0dB}: $\angle A_{LG}$ below $f_{0dB} \neq$ Important

→ As long as $\angle A_{LG}$ recovers

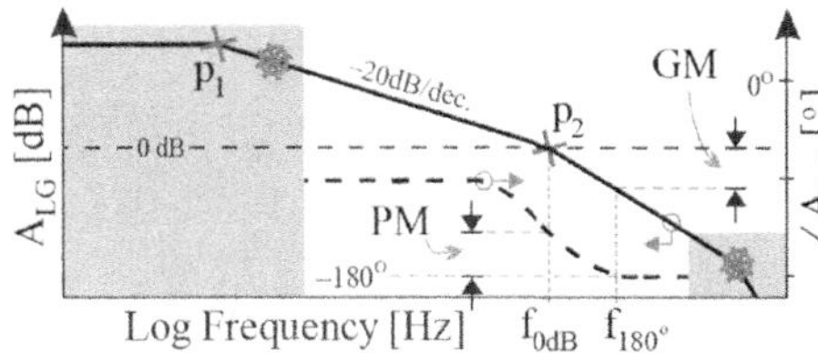

Power-Up: A_{LG} & f_{0dB} rise with v_{DD} power

→ f_{0dB} shifts across f_O ∴ Keep $\angle A_{LG}$ below 180° up to $10f_{0dB}$

Strategy: p_1 reduces A_{LG} to 0 dB

Cancel intermediate poles with zeros

$p_2 \approx f_{0dB}$ → $PM \approx 180° - 90°$ (from p_1) $- 45°$ (from p_2) $= 45°$

Out-of-phase zeros invert & add gain → Keep over $10f_{0dB}$

E. Gain–Bandwidth Product ≡ GBW

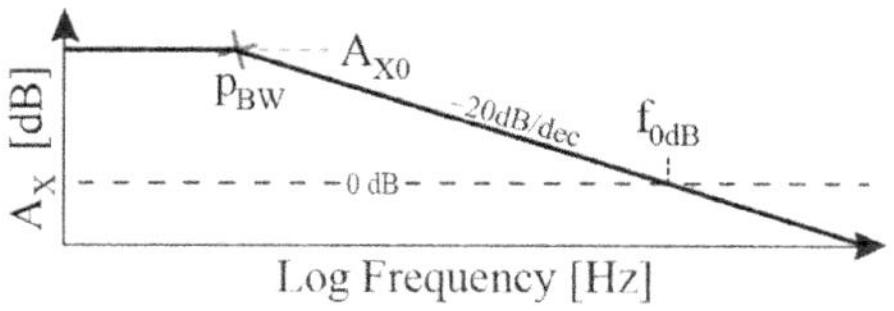

Condition: Drop in gain past 1 pole

p_{BW} ≡ Bandwidth-Setting Pole ≡ –3-dB f_O

f_{0dB} ≡ Unity-gain frequency

Past p_{BW}: $GBW \equiv A_{X0}p_{BW} \approx A_X f_O\big|_{f_O \geq p_{BW}} = (1)f_{0dB} = \text{Constant}$

A_X falls 10× every 10× rise in f_O → Fall in A_X cancels rise in f_O

∴ Use GBW to project A_X & f_{0dB} past p_{BW} ←

$$\text{General: } A_X = A_{X0}\left[\frac{(1+s/2\pi z_1)\ldots(1+s/2\pi z_A)\ldots}{(1+s/2\pi p_1)\ldots(1+s/2\pi p_A)\ldots}\right] \approx A_{X0}\left(\frac{f_O p_1 p_2 \ldots}{z_1 f_O f_O \ldots}\right) = A_{X0}\left(\frac{p_1 \ldots p_N}{z_1 \ldots f_O}\right)$$

$s = i(2\pi f_O)$ $\qquad z_1, p_1, p_2 \ldots \ll f_O \ll z_A, p_A, p_B \ldots$

F. Loop Variations

Pre-Amplifier: Amplifies A_{FW} & $\frac{1}{\beta_{FB}}$

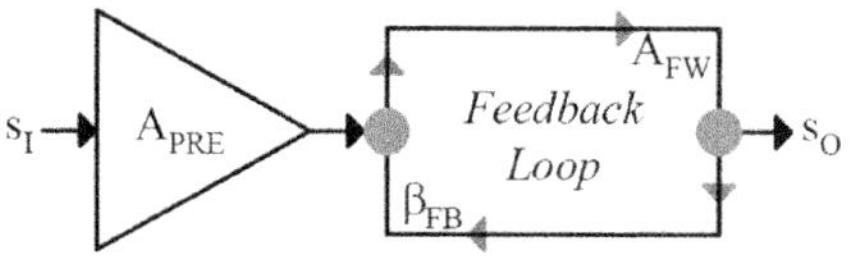

Fwd Gain ↑

$$A_X = A_{PRE}\left(A_{FW} \,||\, \frac{1}{\beta_{FB}}\right) = (A_{PRE}A_{FW}) \,||\, \frac{A_{PRE}}{\beta_{FB}} \equiv A_F \,||\, A_\beta$$

↓ FB Gain

Multiple Taps: Superimpose $s_I A_{CL}$'s

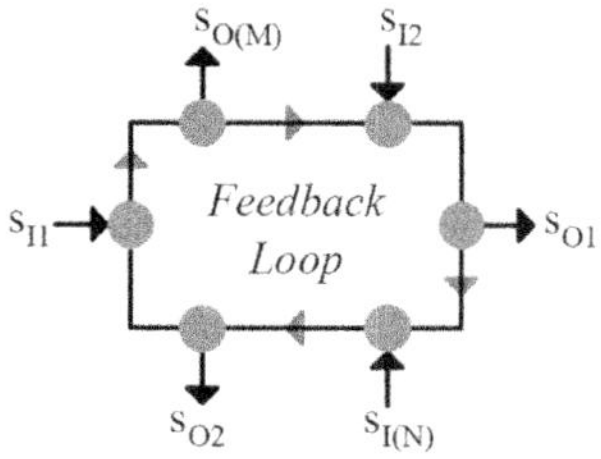

$$s_{O(X)} = \sum_{K=1}^{N} s_{I(K)} A_{CL(K)} = \sum_{K=1}^{N} s_{I(K)}\left(A_{FW(K)} \,||\, \frac{1}{\beta_{FB(K)}}\right)$$

Parallel Paths: Highest translations dominate

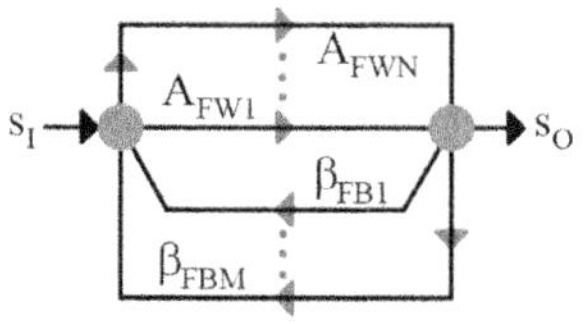

$$A_{CL} = \Sigma A_{FW(X)} \,||\, \frac{1}{\Sigma\beta_{FB(X)}} \approx \text{Max}\{A_{FW(X)}\} \,||\, \frac{1}{\text{Max}\{\beta_{FB(X)}\}}$$

Embedded Loops: Product of stable closed-loop translations

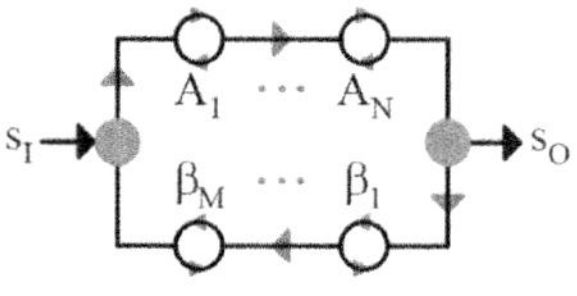

$$A_{CL} = A_{FW} \,||\, \frac{1}{\beta_{FB}} = \prod_{X=1}^{N} A_X \,||\, \frac{1}{\prod_{X=1}^{M} \beta_X}$$

6.2. Op-Amp Translations

Op Amp ≡ Low-R_O Diff. Amp

→ v_O source

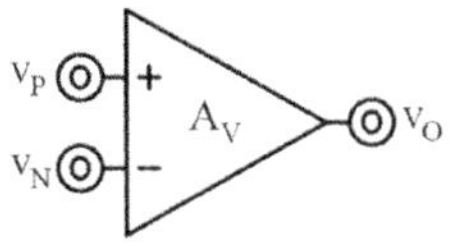

$$A_V \equiv \frac{v_O}{v_P - v_N} \approx \frac{A_{V0}}{1 + s/2\pi p_A}$$

OTA ≡ High-R_O Diff. Amp

→ i_O source

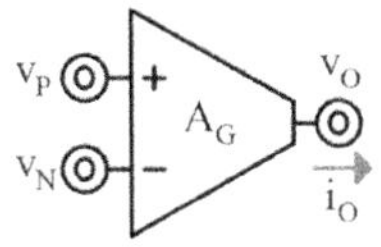

$$A_G \equiv \frac{i_O}{v_P - v_N} \approx A_{G0}$$

Non-Inverting Op Amp: $v_O = (v_{IN} - v_{FB})A_V$ → A_V mixes v_{IN} & v_{FB}

$$A_{FW} \equiv \frac{v_O}{v_E} = \frac{v_O}{v_{IN} - v_{FB}} \approx A_V \qquad \beta_{FB} \equiv \frac{v_{FB}}{v_O} \approx \frac{R_1}{R_1 + R_2} \qquad A_{LG} = A_{FW}\beta_{FB}$$

$$f_{0dB} \approx A_{LG0}p_A = A_{FW0}\beta_{FB}p_A \approx A_{V0}\left(\frac{R_1}{R_1 + R_2}\right)p_A$$

$$A_{VO} = A_{FW} \,||\, \frac{1}{\beta_{FB}} \approx \left(A_{V0} \,||\, \frac{R_1 + R_2}{R_1}\right)\left(\frac{1}{1 + s/2\pi f_{0dB}}\right)$$

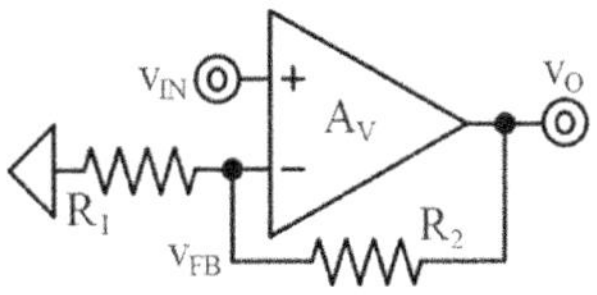

Inverting Op Amp: $v_N = (i_I - i_{FB})(R_1 \,||\, R_2)$ → v_N mixes i_I & i_{FB}

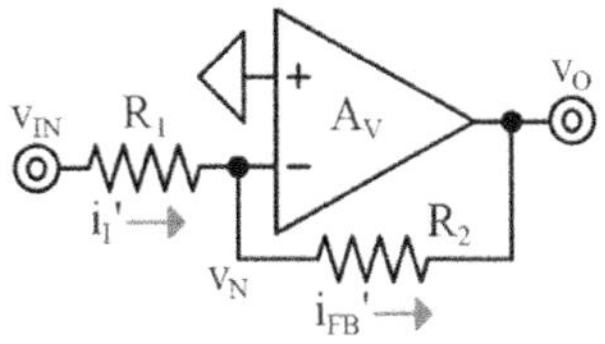

A_V mixes 0 & v_N

But not v_{IN}

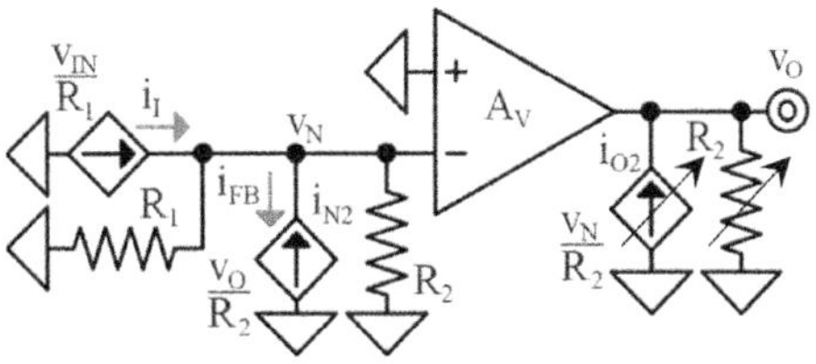

$$A_{FW} \equiv \frac{v_O}{i_E} = \frac{v_O}{i_I - i_{FB}} \approx (R_1 \,||\, R_2)\left(\frac{-A_{V0}}{1 + s/2\pi p_A}\right) \qquad \beta_{FB} \equiv \frac{i_{FB}}{v_O} = -\frac{i_{N2}}{v_O} = -\frac{1}{R_2}$$

$$A_{LG} = A_{FW}\beta_{FB} \approx \left(\frac{R_1}{R_1 + R_2}\right)\left(\frac{A_{V0}}{1 + s/2\pi p_A}\right)$$ → Same A_{LG}

With good op amps: R_2 effects on v_O are negligible

$$f_{0dB} \approx A_{LG0}p_A = A_{FW0}\beta_{FB}p_A \approx \left(\frac{R_1}{R_1 + R_2}\right)A_{V0}p_A$$ → Same f_{0dB}

$$A_{VO} \equiv \frac{v_O}{v_{IN}} = \left(\frac{i_I}{v_{IN}}\right)\left(A_{FW} \,||\, \frac{1}{\beta_{FB}}\right) \approx -\left(\frac{R_2 A_{V0}}{R_1 + R_2} \,||\, \frac{R_2}{R_1}\right)\left(\frac{1}{1 + s/2\pi f_{0dB}}\right)$$ → Lower $A_{F/\beta}$

6.3. Stabilizers: A. Strategies

Type I: Dominant Pole

p_{S1} reduces A_{LG} towards 0 dB

p_{S2} can be near f_{0dB}

PM ≈ 180° – 90° – 45° = 45°

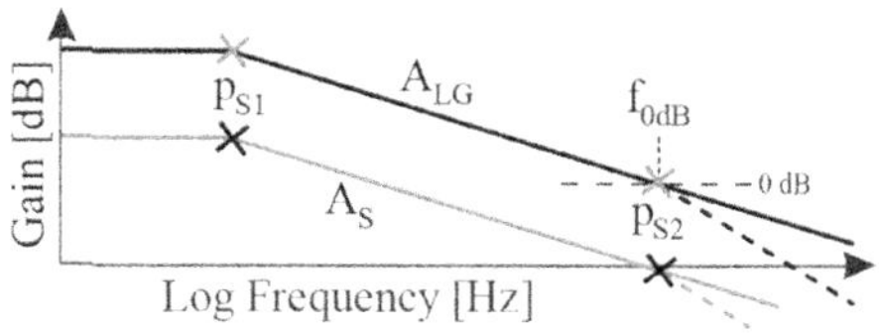

Type II: Pole–Zero Pair

z_{S1} recovers p_1's phase

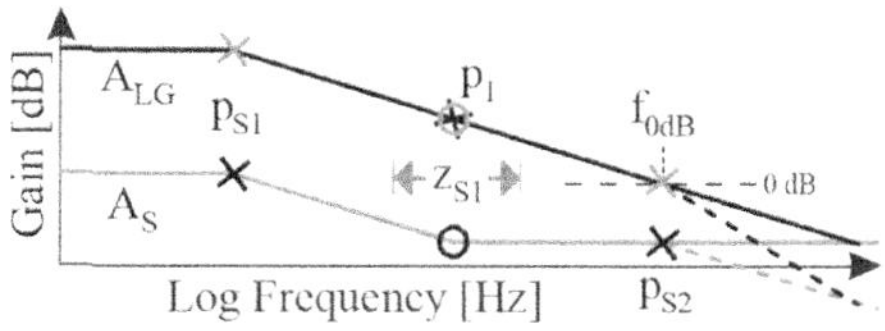

Type III: Pole–Zero–Zero Triplet

z_{S2} recovers p_2's phase

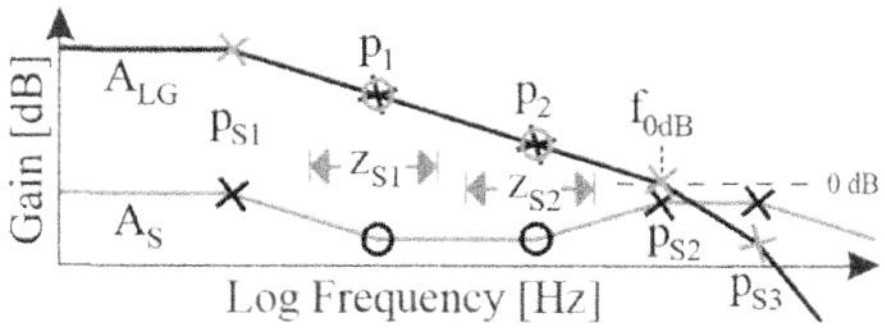

B. Amplifier Translations

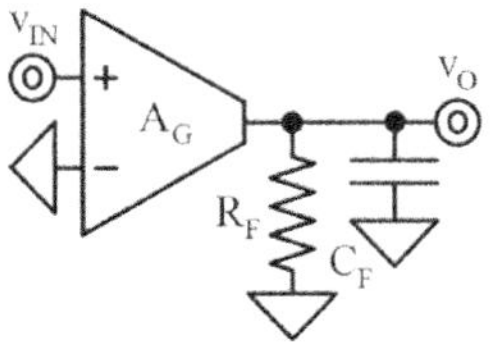

C_F shunts R_F → p_F ∴ $A_S \approx \dfrac{A_G R_F}{1+s/2\pi p_F}$

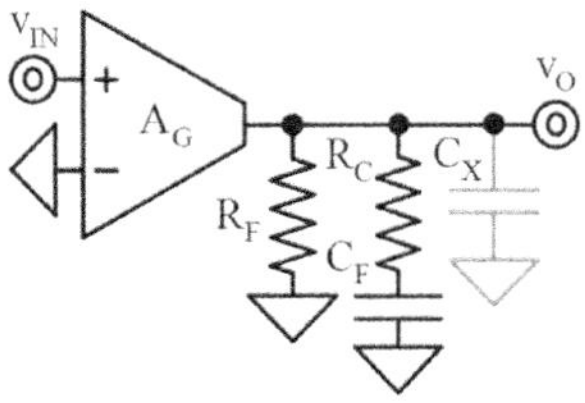

C_F & R_C shunt R_F → p_C

R_C current-limits C_F → z_{CX}

C_X shunts $R_C \parallel R_F$ → p_O

$$A_S \approx \frac{A_G R_F\left(1+s/2\pi z_{CX}\right)}{\left(1+s/2\pi p_C\right)\left(1+s/2\pi p_O\right)}$$

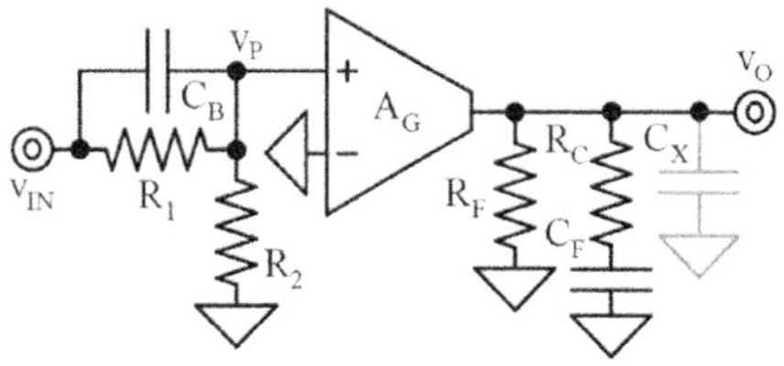

C_B bypasses R_1 → z_B

$R_1 \parallel R_2$ current-limits C_B → p_{BX}

$$\therefore \quad A_S \approx \left(\frac{R_2 A_G R_F}{R_1+R_2}\right)\left[\frac{\left(1+s/2\pi z_B\right)\left(1+s/2\pi z_{CX}\right)}{\left(1+s/2\pi p_{BX}\right)\left(1+s/2\pi p_C\right)\left(1+s/2\pi p_O\right)}\right]$$

Example: Determine R_2, C_B, A_G, C_F, R_C, & C_X when $A_{S0} = 40$ V/V, $p_{S1} = 1$ kHz, $z_{S1} = z_{S2} = 10$ kHz, $p_{S2} \geq 100$ kHz, $p_{S3} \geq 1$ MHz, $R_1 = R_F = 500$ kΩ.

Solution:

$$\frac{z_{S1}}{p_{S1}} \equiv \frac{z_{CX}}{p_C} \approx \frac{R_C + R_F}{R_C} = \frac{R_C + 500k}{R_C} \equiv \frac{10\ \text{kHz}}{1\ \text{kHz}} \qquad \therefore\ R_C = 56\ \text{k}\Omega$$

$$\frac{p_{S2}}{z_{S2}} \equiv \frac{p_{BX}}{z_B} \approx \frac{R_1}{R_1 \| R_2} = \frac{500k}{500k \| R_2} \equiv \frac{100\ \text{kHz}}{10\ \text{kHz}} \qquad \therefore\ R_2 = 56\ \text{k}\Omega$$

$$A_{S0} = \frac{R_2 A_G R_F}{R_1 + R_2} = \frac{(56k)A_G(500k)}{500k + 56k} \equiv 40\ \text{V/V} = 32\ \text{dB} \qquad \therefore\ A_G = 790\ \mu\text{A/V}$$

$$p_{S1} \equiv p_C \approx \frac{1}{2\pi(R_C + R_F)C_F} = \frac{1}{2\pi(56k + 500k)C_F} \equiv 1\ \text{kHz} \qquad \therefore\ C_F = 290\ \text{pF}$$

$$z_{S2} \equiv z_B \approx \frac{1}{2\pi R_1 C_B} = \frac{1}{2\pi(500k)C_B} \equiv 10\ \text{kHz} \qquad \therefore\ C_B = 32\ \text{pF}$$

$$p_{S3} = p_O \approx \frac{1}{2\pi(R_C \| R_F)C_X} = \frac{1}{2\pi(56k \| 500k)C_X} \geq 1\ \text{MHz} \qquad \therefore\ C_X \leq 3.2\ \text{pF}$$

C. Feedback Translations

$C_{F2} \ll C_F, C_B$

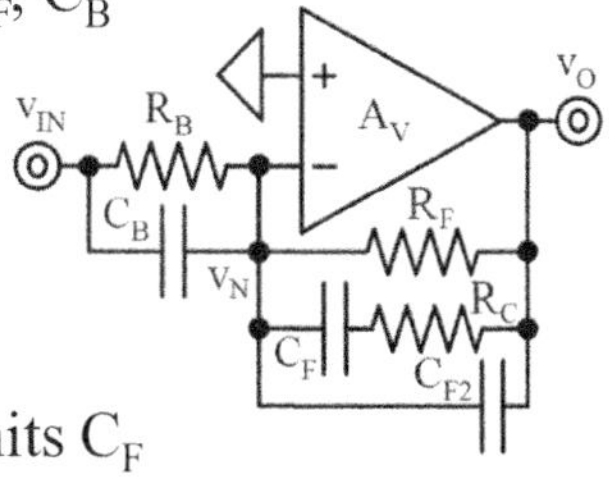

$$A_\beta \approx -\left(\frac{R_F}{R_B}\right)\left[\frac{(1 + s/2\pi z_{CX})(1 + s/2\pi z_B)}{(1 + s/2\pi p_C)(1 + s/2\pi p_{C2})}\right]$$

p_C: C_F & R_C (with C_{F2}) shunt R_F $\qquad$ z_{CX}: R_C current-limits C_F

p_{C2}: C_{F2} shunts $R_C \| R_F$ $\qquad$ z_B: C_B bypasses R_B $\qquad$ W/o p_{C2} → p_X^2

$$A_F \approx \left(\frac{-R_F A_V}{R_B + R_F}\right)\left[\frac{(1 + s/2\pi z_{CX})(1 + s/2\pi z_B)}{(1 + s/2\pi p_C''')(1 + s/2\pi p_{BX}')}\right]$$

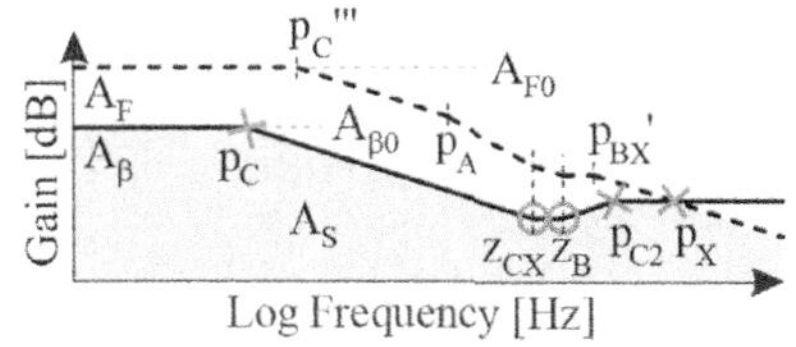

p_C''': C_F & R_C (with C_{F2}) shunt $R_F \| R_B$ $\qquad$ z_{CX}: R_C current-limits C_F

↓

z_B: C_B bypasses R_B $\quad$ < ——→ $\quad$ p_{BX}': $R_C \| R_F \| R_B$ current-limits C_B & C_{F2}

$$A_S \approx A_\beta \text{ until } |A_F| \approx \frac{R_F A_{V0} p_A p_C''' p_{BX}'}{(R_B + R_F) f_O z_{CX} z_B} \leq |A_\beta| \approx \frac{R_F p_C p_{C2}}{R_B z_{CX} z_B} \rightarrow p_X \approx \frac{R_B A_{V0} p_A p_C''' p_{BX}'}{(R_B + R_F) p_C p_{C2}}$$

Example: Determine p_A, R_B, C_B, C_F, R_C, & C_{F2} with $|A_{S0}|$, p_{S1}, z_{S1}, z_{S2}, p_{S2}, p_{S3}, R_F from earlier example when A_{V0} is 1 kV/V.

Solution: Same p_{S1}, z_{S1} $\therefore$ $R_C = 56\ k\Omega$, $C_F = 290$ pF from earlier example

$$A_{S0} \approx -\left(\frac{R_F A_{V0}}{R_B + R_F} \parallel \frac{R_F}{R_B}\right) = -\left(\frac{(500k)(1k)}{R_B + (500k)} \parallel \frac{500k}{R_B}\right) \equiv -40\ \text{V/V} \quad \therefore \quad R_B = 12\ k\Omega$$

$$z_{S2} \equiv z_B \approx \frac{1}{2\pi R_B C_B} = \frac{1}{2\pi(12k)C_B} \equiv 10\ \text{kHz} \quad \therefore \quad C_B = 1.3\ \text{nF}$$

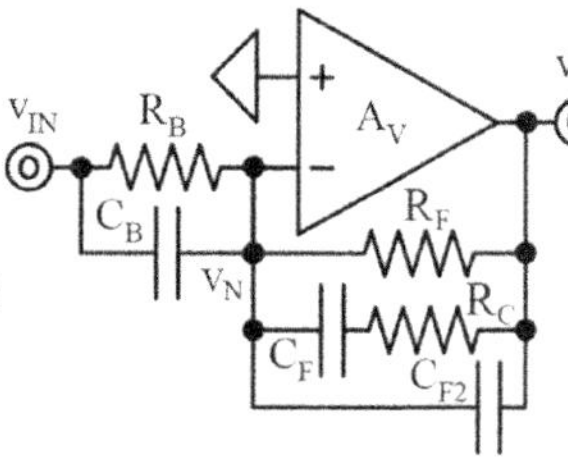

$$p_{S2} \equiv p_{C2} \approx \frac{1}{2\pi(R_C \parallel R_F)C_{F2}} = \frac{1}{2\pi(56k \parallel 500k)C_{F2}} > 100\ \text{kHz}$$

$\therefore$ $C_{F2} < 32$ pF

$$p_{S3} \equiv p_X \approx \frac{\overbrace{A_{V0}p_A\left[R_B/(R_B+R_F)\right](R_C+R_F)}^{\text{From } p_C}\overbrace{(R_F \parallel R_C)C_{F2}}^{\text{From } p_{C2}}}{\underbrace{\left[R_C+(R_B \parallel R_F)\right]}_{\text{From } p_C'''}\underbrace{(R_F \parallel R_C \parallel R_B)(C_B+C_{F2})}_{\text{From } p_{BX}'}} = (24m)A_{V0}p_A > 1\ \text{MHz}$$

$\therefore$ $p_A > 42$ kHz

SPICE Simulation

* Type-III Inverting Feedback Translation

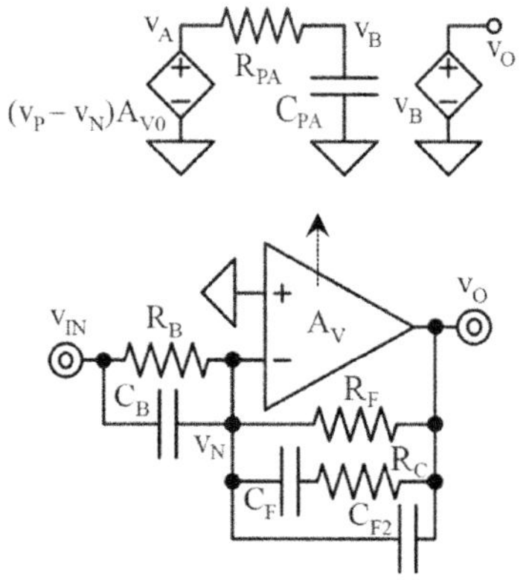


```
vin vin 0 dc=0 ac=1
rb vin vn 12k
cb vin vn 1.3n
eav0 va 0 0 vn 1000
rpa va vb 1
cpa vb 0 3.791u
eavo vo 0 vb 0 1
```

(rpa, cpa: $p_A \equiv 42$ kHz)

eavo: → Buffer (unload) p_A

```
rf vn vo 500k
cf vn vx 290p
rc vx vo 56k
cf2 vn vo 32p
.ac dec 1000 10 10e6
.end
```

* Plot v_O in dB

* Voltage-controlled voltage source: ex vop von vcp vcn gain

D. Mixed Translations

Assume p_A >> Others

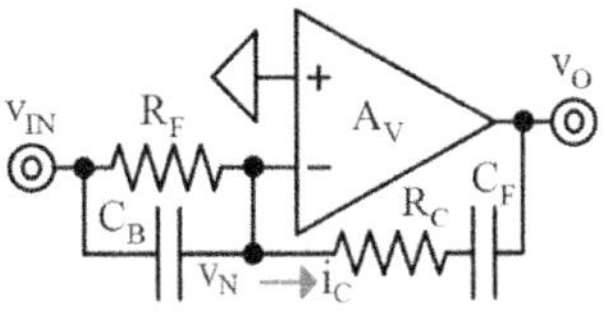

$$A_F \approx (1)\left[\frac{(1+s/2\pi z_{CX})(1+s/2\pi z_B)}{(1+s/2\pi p_C)(1+s/2\pi p_{BX}")}\right]\left(\frac{-A_{V0}}{1+s/2\pi p_A}\right)$$

p_C: C_F & R_C (with C_B) shunt R_F

z_{CX}: R_C current-limits C_F

z_B: C_B bypasses R_F

p_{BX}": $R_C \parallel R_F$ current-limits C_B

$$A_\beta \approx -\left(\frac{Z_C}{R_F}\right)\left(1+\frac{s}{2\pi z_{CX}}\right)\left(1+\frac{s}{2\pi z_B}\right)$$

$1 \div s/2\pi p_F = p_F/f_O i$

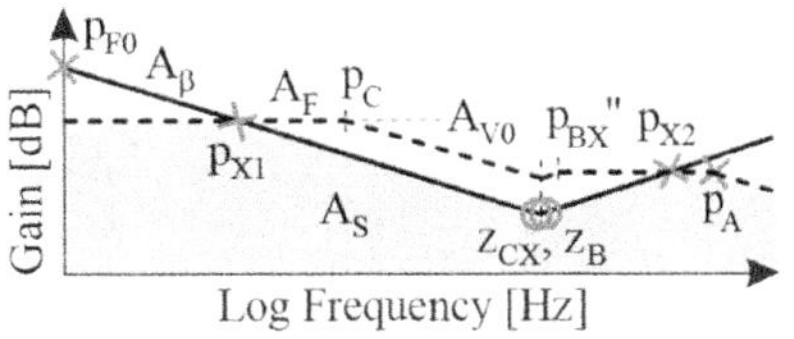

p_{F0}: A_β falls as C_F shorts

z_{CX}: R_C current-limits C_F

z_B: C_B bypasses R_F

$A_S \approx A_{F0}$ until $|A_\beta| \approx \frac{p_F}{f_O} \le |A_{F0}| = A_{V0}$ → $p_{X1} \approx \frac{p_F}{A_{V0}} \approx \frac{1}{2\pi R_F C_F A_{V0}}$

$A_S \approx A_\beta$ until $|A_\beta| \approx \frac{p_F f_O}{z_{CX} z_B} \ge |A_F| \approx A_{V0}\left(\frac{p_C p_{BX}"}{z_{CX} z_B}\right)$ → $p_{X2} \approx A_{V0}\left(\frac{p_C p_{BX}"}{p_F}\right)$

Example: Determine C_B, A_{V0}, p_A, C_F, R_C, & z_{S2} with $|A_{S0}|$, p_{S1}, z_{S1}, p_{S2}, p_{S3} from earlier example.

Solution: A_{V0} magnifies C_F's shunting effect.

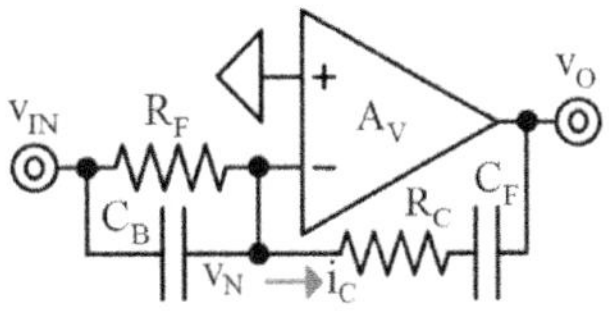

$A_{S0} \approx -A_{V0} \equiv -40$ V/V = 32 dB

$$p_{S1} \equiv p_{X1} \approx \frac{p_F}{A_{V0}} = \frac{1}{2\pi R_F C_F A_{V0}} = \frac{1}{2\pi(500k)C_F(40)} \equiv 1\ \text{kHz}$$

$\therefore\ C_F = 8.0$ pF

$$z_{S1} \equiv z_{CX} \approx \frac{1}{2\pi R_C C_F} = \frac{1}{2\pi R_C (8.0p)} \equiv 10\ \text{kHz}$$

$\therefore\ R_C = 2.0$ MΩ

$$p_{S2} \equiv p_{X2} \approx \frac{A_{V0} p_C p_{BX}"}{p_F} = \frac{A_{V0}(2\pi)R_F C_F}{2\pi(R_C + R_F)C_F(2\pi)(R_C \parallel R_F)C_B} = \frac{A_{V0}}{2\pi R_C C_B} > 100\ \text{kHz}$$

$\therefore\ C_B < 32$ pF

$$z_{S2} \equiv z_B \approx \frac{1}{2\pi R_F C_B} > \frac{1}{2\pi(500k)(32p)} = 10\ \text{kHz}$$

$p_{S3} \approx p_A > 1$ MHz

6.4. Voltage Control: A. Controller

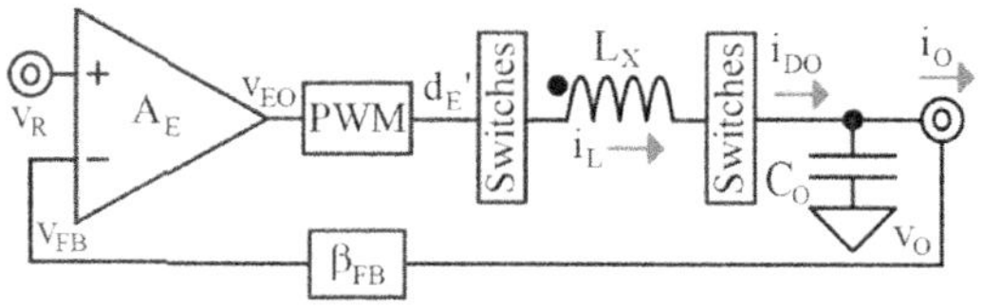

Objective: Supply i_O needed to set & keep v_O steady

Controller: Scale v_O to v_{FB} ≡ Feedback Scaler β_{FB}

Sense & amplify v_E & stabilize loop ≡ Error Amp A_E

Translate v_{EO} to d_E' ≡ Pulse-Width Modulator (PWM)

Power Stage: Transfer P_{IN} to v_O ≡ Switched Inductor (SL)

Supply/absorb $i_L - i_O$ mismatch ≡ Output Capacitor C_O

Feedback Action: A_E senses & amplifies the v_E that adjusts d_E' & i_L

$$\text{So} \quad v_{FB} \approx v_R \quad \therefore \quad v_O = \frac{v_{FB}}{\beta_{FB}} \approx \frac{v_R}{\beta_{FB}}$$

Feedback Objectives:

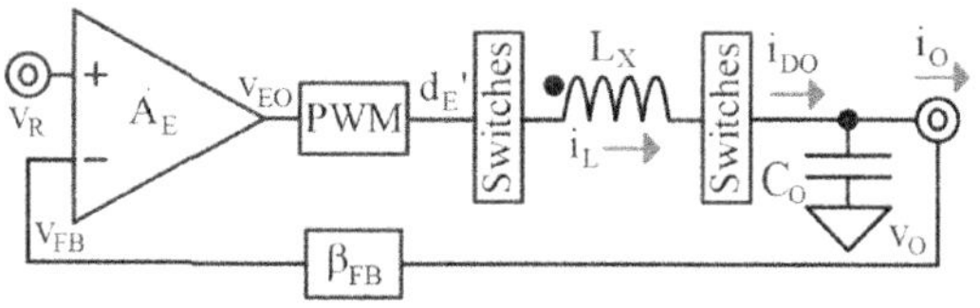

Accurate → High A_{LG}

Fast Response → High f_{0dB}

B. Loop Gain

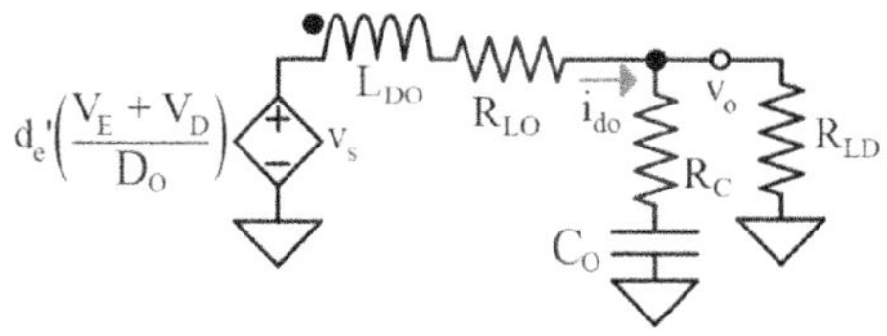

$$A_{LG} \equiv \frac{v_{FB}}{v_E} = A_E A_{PWM} A_{SL} \beta_{FB} \qquad \beta_{FB} \equiv \frac{v_{FB}}{v_O} \approx \frac{v_R}{v_O}$$

$$A_{PWM0} \equiv \frac{d_e'}{v_{eo}} \approx \frac{\Delta d_{E(MAX)}'}{\Delta v_{EO}'} \approx \frac{1}{v_{EO(PP)}}$$ → Δv_{EO} needed to sweep d_E' from 0 to 1

PWM delays v_{eo}:d_e' translations → A_{PWM0} up to p_{SW} $\therefore$ $A_{PWM} = \dfrac{A_{PWM0}}{1 + s/2\pi p_{SW}}$

$$A_{SL(CCM)} \equiv \frac{v_o}{d_e'} \approx \left(\frac{V_E + V_D}{D_O}\right)\left(\frac{R_{LD}}{R_{LO} + R_{LD}}\right)\left\{\frac{(1 + s/2\pi z_C)(1 - s/2\pi z_{DO})}{\left[(s/2\pi p_{LC})^2 + s/2\pi p_{LC} Q_{LC} + 1\right]}\right\}$$

Usually, R_C = Very low → $z_C >> p_{LC}$ And $z_{DO} > p_{LC}$, but often not by much

C. Voltage Mode

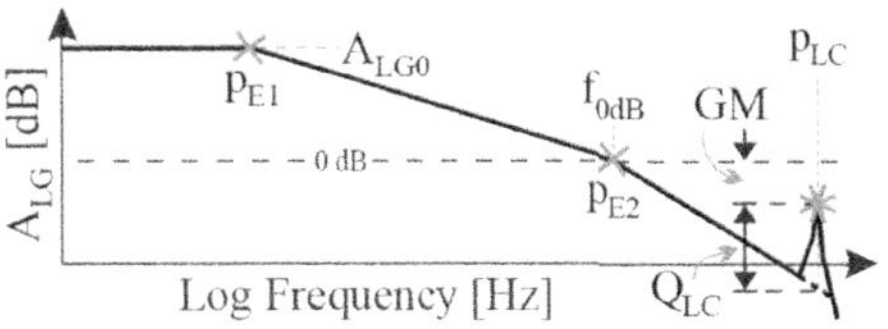

Type I: Dominant Pole

p_{E1} reduces A_{LG} to 0 dB so $f_{0dB} \le \frac{p_{LC}}{10}$

$p_{E2} \equiv f_{0dB} \approx A_{LG0}p_{E1}$ suppresses LC peak → PM ≈ 45° GM ≤ 40 dB

$$A_{LG}\Big|_{f_{LC}>p_{E1},p_{E2}} \approx \left(\frac{A_{LG0}p_{E1}p_{E2}}{f_{LC}^2}\right)Q_{LC} \approx \left(\frac{f_{0dB}^2}{p_{LC}^2}\right)\left[\frac{1}{2\pi(R_{LO}+R_C)C_Op_{LC}}\right] = \frac{1}{GM}$$

$$Q_{LC} \to \frac{f_{CS}}{f_{LC}}$$

Type III: Pole–Zero–Zero Triplet

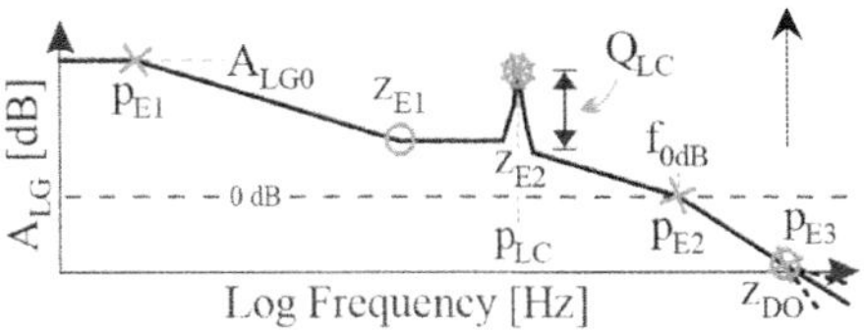

z_{E1} & z_{E2} recover phase lost to $p_{LC} < f_{0dB}$

$p_{E2} \equiv f_{0dB}$ → Loses 45° at f_{0dB}

$p_{E3} \ge 10f_{0dB}$ → Loses ≤ 6°

$z_{DO} \ge 10f_{0dB}$ → Loses ≤ 6°

$$A_{LG}\Big|_{f_{0dB}>p_{E1},z_{E1},z_{E2},p_{LC}} \approx \frac{A_{LG0}p_{E1}p_{LC}^2}{z_{E1}z_{E2}f_{0dB}} = 1 \quad \therefore \quad f_{0dB} \approx A_{LG0}\left(\frac{p_{E1}p_{LC}^2}{z_{E1}z_{E2}}\right)$$

Example: Determine A_{E0}, p_{E1}, & PM so A_{LG0} is 100 V/V when $z_{E1} = z_{E2} = 10\%f_{0dB}$, $p_{E2} \approx f_{0dB} = 100$ kHz, $p_{E3} = f_{SW} = 1$ MHz, $v_O = 4$ V, $v_R = 1$ V, $v_E = 2$ V, $v_D = 4$ V, $d_O = 1$, $v_{EO(PP)} = 500$ mV, $L_X = 10$ µH, $R_L = 50$ mΩ, $C_O = 5$ µF, $R_C = 10$ mΩ.

Solution:

$$\beta_{FB} \approx \frac{v_R}{v_O} = \frac{1}{4} = 250 \text{ mV/V} \qquad A_{PWM0} \approx \frac{1}{v_{EO(PP)}} = \frac{1}{500m} = 2 \text{ V}^{-1}$$

$$A_{SL0} \approx V_E + V_D = 2 + 4 = 6 \text{ V} \qquad z_C = \frac{1}{2\pi R_C C_O} = \frac{1}{2\pi(10m)(5\mu)} = 3.2 \text{ MHz}$$

$$A_{LG0} = A_{E0}A_{PWM0}A_{SL0}\beta_{FB} \approx A_{E0}(2)(6)(250m) \equiv 100 \text{ V/V} \quad \therefore \quad A_{E0} = 33 \text{ V/V}$$

$d_O = 1$ ∴ $L_{DO} = L_X$ $R_{LO} = R_L$ No z_{DO}

$$p_{LC} = \frac{1}{2\pi\sqrt{L_X C_O}} = \frac{1}{2\pi\sqrt{(10\mu)(5\mu)}} = 22\ \text{kHz}$$

$$f_{0dB} \approx A_{LG0}\left(\frac{p_{E1}p_{LC}^{\ 2}}{z_{E1}z_{E2}}\right) = 100\left[\frac{p_{E1}(22k)^2}{(10k)(10k)}\right] \equiv 100\ \text{kHz} \quad \therefore \quad p_{E1} = 210\ \text{Hz}$$

$$p_{LC} = 22\ \text{kHz} < f_{0dB} = 100\ \text{kHz} \quad \therefore \quad \angle A_{LC(0dB)} \equiv \angle A_{LC} \text{ at } f_{0dB} \approx -180°$$

$$PM = 180° - \tan^{-1}\left(\frac{f_{0dB}}{p_{E1}}\right) + \tan^{-1}\left(\frac{f_{0dB}}{z_{E1}}\right) + \tan^{-1}\left(\frac{f_{0dB}}{z_{E2}}\right) + \angle A_{LC(0dB)}$$

$$- \tan^{-1}\left(\frac{f_{0dB}}{p_{E2}}\right) - \tan^{-1}\left(\frac{f_{0dB}}{p_{E3}}\right) - \tan^{-1}\left(\frac{f_{0dB}}{p_{SW}}\right) + \tan^{-1}\left(\frac{f_{0dB}}{z_C}\right)$$

$$\approx 180° - 90° + 84° + 84° - 180° - 45° - 6° - 6° + 2° = 23°$$

D. Current Mode

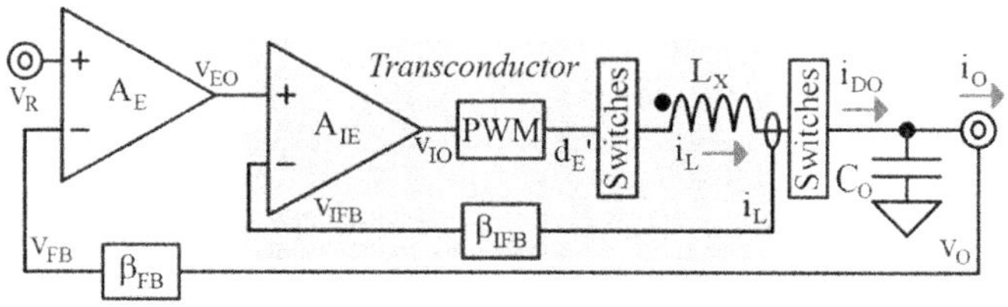

Current Loop:

A_{IE} senses & amplifies the v_{IE} that adjusts d_E' & i_L so $v_{IFB} \approx v_{EO}$.

A_E senses & amplifies the v_E that adjusts v_{EO} so $v_{FB} \approx v_R$.

Note: A_{ILG} to i_L excludes z_{DO}.

$\therefore \quad i_L = \frac{v_{IFB}}{\beta_{IFB}} \approx \frac{v_{EO}}{\beta_{IFB}}$ up to $f_{I0dB} = p_G$ → $i_L \neq f(sL_X)$ $\therefore$ No p_L in p_{LC}

$$\rightarrow \quad A_G \equiv \frac{i_L}{v_{EO}} \approx \frac{1/\beta_{IFB}}{(1+s/2\pi p_G)(1+s/2\pi p_{SW})}$$

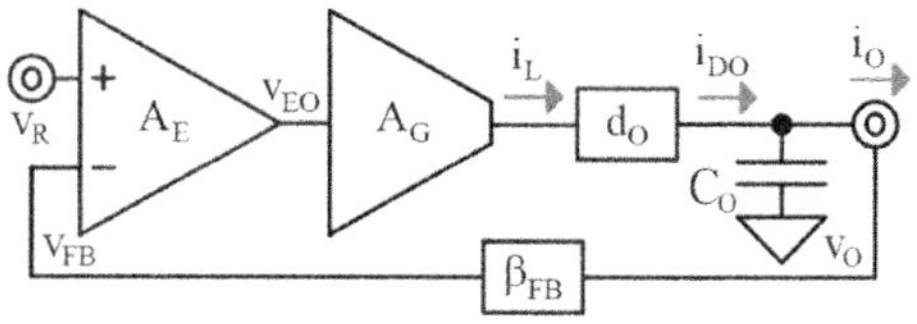

Loop Gain:

$$A_{LG} \equiv \frac{v_{FB}}{v_E} = A_E A_G D_O[(Z_C + R_C) \parallel R_{LD}]\beta_{FB} \approx \frac{A_E A_{G0} D_O R_{LD} \beta_{FB}(1+s/2\pi z_C)(1+s/2\pi z_{DO})}{(1+s/2\pi p_G)(1+s/2\pi p_{SW})(1+s/2\pi p_{CP})}$$

Usually, $p_{CP} < p_{LC} < p_G$ → Current loop effectively splits p_{LC} into p_{CP} & p_G.

Type I: Dominant Pole

p_{CP} reduces A_{LG} to $f_{0dB} \le p_G \le \frac{p_{E1}}{10}, \frac{f_{SW}}{10}$ → Inherently stable

$R_C << R_{LD}$ → $R_C + R_{LD} \approx R_{LD}$ ∴ R_{LD} in A_{LG0} & p_{CP} cancels in f_{0dB}:

$$f_{0dB} \approx A_{LG0}p_{CP} \approx \frac{A_{E0}A_{G0}D_O\beta_{FB}}{2\pi C_O} \approx \frac{A_{E0}D_O\beta_{FB}}{2\pi\beta_{IFB}C_O}$$

Type II: Pole–Zero Pair

p_{E1}, z_{E1}, p_{CP} project A_{LG} to $f_{0dB} \le p_G \le \frac{p_{E2}}{10}, \frac{f_{SW}}{10}$

$$A_{LG}\Big|_{f_{0dB} > p_{E1}, z_{E1}, p_{CP}} \approx \frac{A_{LG0}p_{E1}p_{CP}}{z_{E1}f_{0dB}} = 1 \quad \rightarrow \quad f_{0dB} \approx A_{LG0}\left(\frac{p_{E1}p_{CP}}{z_{E1}}\right)$$

Note: $p_{CP} < p_{LC} < f_{0dB} \approx p_G$ → Current mode can be faster than voltage mode.

Example: Determine A_{E0}, p_{E1}, p_{CP}, & PM so $f_{0dB} = p_G = 100$ kHz when $\beta_{IFB} = 1\ \Omega$, $d_O = 33\%$ with other parameters from earlier example.

Solution: $\beta_{FB} \approx 250$ mV/V, $f_{SW} = 1$ MHz, $z_C = 3.2$ MHz from earlier example

$$p_{CP} \approx \frac{1}{2\pi(R_C + R_{LD})C_O} = \frac{1}{2\pi(10m + 100)(5\mu)} = 320\ \text{Hz}$$

$$f_{0dB} \approx \frac{A_{E0}A_{G0}D_O\beta_{FB}}{2\pi C_O} \approx \frac{A_{E0}D_O\beta_{FB}}{2\pi C_O\beta_{IFB}} = \frac{A_{E0}(33\%)(250m)}{2\pi(5\mu)(1)} \equiv p_G = 100\ \text{kHz} << z_C$$

∴ $A_{E0} = 38$ V/V $= 32$ dB $\qquad p_{E1} \ge 10f_{0dB} = 1$ MHz

$$PM = 180° - \tan^{-1}\left(\frac{f_{0dB}}{p_{CP}}\right) - \tan^{-1}\left(\frac{f_{0dB}}{p_G}\right) - \tan^{-1}\left(\frac{f_{0dB}}{p_{E1}}\right) - \tan^{-1}\left(\frac{f_{0dB}}{p_{SW}}\right)$$

$$\approx 180° - 90° - 45° - 6° - 6° = 33°$$

E. Discontinuous Conduction

Switched Inductor: $A_{SL(DCM)} \equiv \dfrac{v_o}{d_e'} \approx \left(\dfrac{I_{L(PK)}}{D_E}\right)\left(R_{DO} \parallel R_{LD}\right)\left(\dfrac{1+s/2\pi z_C}{1+s/2\pi p_{CS}}\right)$

$$R_{DO} \approx 2\left(\frac{L_X}{D_O^{\ 2}}\right)\left(\frac{T_{SW}}{T_C^{\ 2}}\right)+\frac{R_L}{D_O}$$

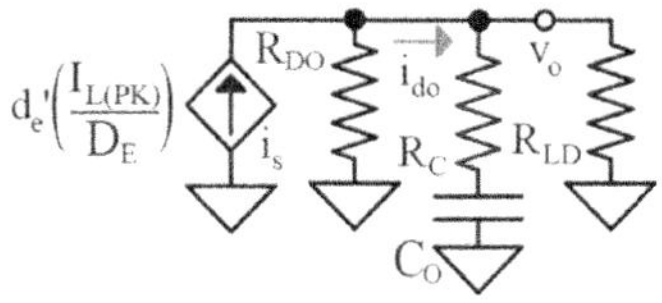

Type I: Dominant Pole

p_{CS} reduces A_{LG} to $f_{0dB} \le p_{E1} \le \dfrac{f_{SW}}{10}$ → Inherently stable

$R_C << R_{DO} \parallel R_{LD}$ → $R_C + (R_{DO} \parallel R_{LD}) \approx R_{DO} \parallel R_{LD}$

∴ $R_{DO} \parallel R_{LD}$ in A_{LG0} & p_{CS} cancels in f_{0dB}:

$$f_{0dB} \approx A_{LG0}p_{CS} = \frac{A_{E0}A_{PWM0}A_{LI}\beta_{FB}}{2\pi C_O} \approx \frac{A_{E0}A_{PWM0}I_{L(PK)}\beta_{FB}}{2\pi C_O D_E}$$

6.5. Current Control: A. Controller

Objective: Supply i_L, i_{DO}, or i_O that is insensitive to v_O.

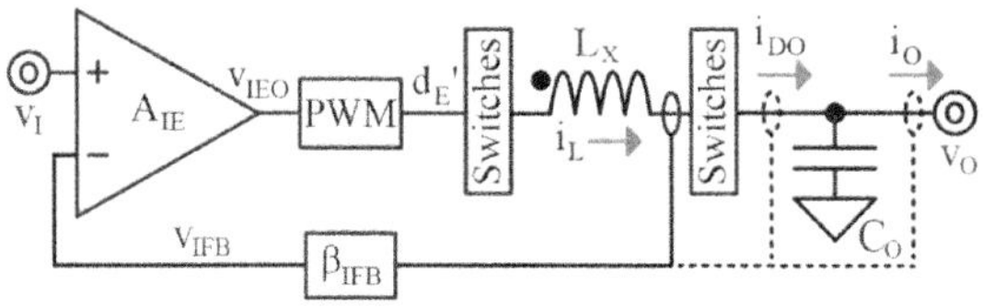

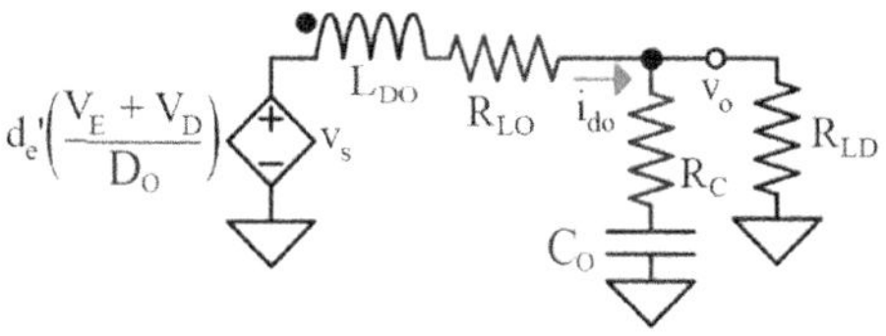

Feedback: A_{IE} senses & amplifies the v_{IE} that adjusts d_E' & $i_{L/DO/O}$

So $v_{IFB} \approx v_I$ ∴ $i_{L/DO/O} = \dfrac{v_{IFB}}{\beta_{IFB}} \approx \dfrac{v_I}{\beta_{IFB}}$

B. i_L Gain: $A_{IL} \equiv \dfrac{i_l}{d_e'} = \left.\dfrac{i_{do}/D_O}{d_e'}\right|_{\text{No } z_{DO}}$

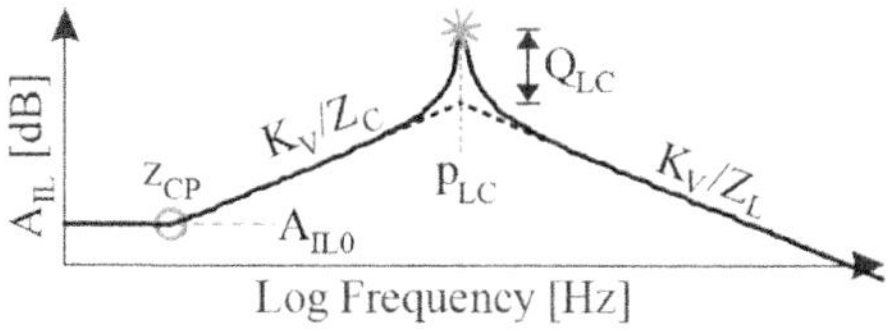

i_L climbs when C_O & R_C shunt R_{LD}

$$A_{IL(CCM)} = \frac{A_{LV(CCM.No\ z_{DO})}/D_O}{Z_{DO}+\left[\left(Z_C+R_C\right)\parallel R_{LD}\right]} \approx \frac{\left(V_E+V_D\right)\left(1+s/2\pi z_{CP}\right)}{\left[D_O^{\ 2}\left(R_{LO}+R_{LD}\right)\right]\left[\left(s/2\pi p_{LC}\right)^2+s/\left(2\pi p_{LC}Q_{LC}\right)+1\right]}$$

C. Type I: Dominant Pole

A_{IE0} increases A_{ILG} so $f_{I0dB} \le p_{IE1}, \dfrac{f_{SW}}{10}$ → Inherently stable

$$A_{ILG}\big|_{f_{I0dB} > p_{LC} > z_{CP}} \approx \frac{A_{ILG0}\, p_{LC}^{\,2}}{z_{CP} f_{I0dB}} = 1 \quad \rightarrow \quad f_{I0dB} \approx A_{ILG0}\left(\frac{p_{LC}^{\,2}}{z_{CP}}\right) = f_{IBW(CL)} = p_G$$

A_{ILG} rises & falls 20 dB/dec. to 0 dB ∴ Stable Low A_{ILG0}

$$A_G \equiv \frac{i_L}{v_I} = A_{IF} \parallel A_{I\beta} = \left(A_{IE} A_{PWM} A_{IL}\right) \parallel \frac{1}{\beta_{IFB}}$$

$$\left|A_{IF}\right|_{z_{CP} < f_O < p_{LC}} = A_{IE0} A_{PWM0} A_{IL0}\left(\frac{f_O}{z_{CP}}\right) \ge |A_{I\beta}| = \frac{1}{\beta_{IFB}} \quad \rightarrow \quad f_O \ge \frac{z_{CP}}{A_{ILG0}} \approx p_{X1}$$

$$A_G \approx \frac{A_{IE} A_{PWM0} A_{IL0}\left(1 + s/2\pi z_{CP}\right)}{\left(1 + s/2\pi p_{X1}\right)\left(1 + s/2\pi p_G\right)\left(1 + s/2\pi p_{SW}\right)}$$

$$\approx \frac{1/\beta_{IFB}}{\left(1 + s/2\pi p_G\right)\left(1 + s/2\pi p_{IE1}\right)\left(1 + s/2\pi p_{SW}\right)} \quad \text{Above } p_{X1}$$

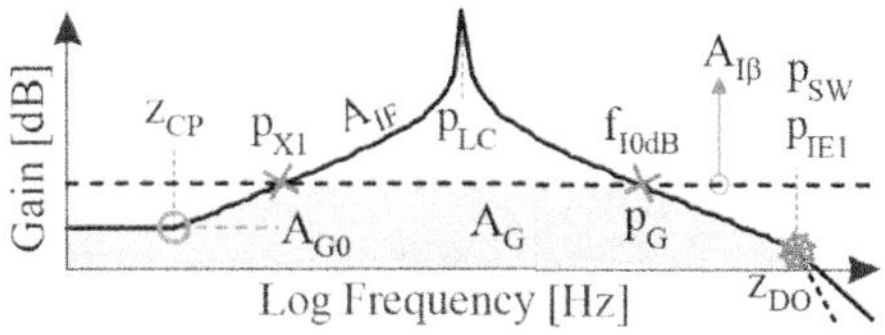

D. Type II: Pole–Zero Pair

A_{E0} increases A_{ILG} so $f_{I0dB} = f_{IBW(CL)} = p_G \le p_{IE2}, \dfrac{f_{SW}}{10}$

p_{IE1} at or below z_{CP} counters z_{CP}

z_{IE1} near p_{LC} recovers phase lost to p_{IE1} ∴ $A_{IF0} > A_{I\beta}$ → Higher A_{ILG0}

$$A_G \approx A_{I\beta} \text{ up to } p_G \quad \rightarrow \quad A_G \approx \frac{1/\beta_{IFB}}{\left(1 + s/2\pi p_G\right)\left(1 + s/2\pi p_{IE2}\right)\left(1 + s/2\pi p_{SW}\right)}$$

$$\left|A_{IF}\right|_{f_O > p_{LC} \ge z_{IE1} > z_{CP} \ge p_{IE1}} \approx A_{IE0} A_{PWM0} A_{IL0}\left(\frac{p_{IE1} p_{LC}^{\,2}}{z_{IE1} z_{CP} f_O}\right) \le |A_{I\beta}| = \frac{1}{\beta_{IFB}}$$

$$\rightarrow \quad f_O \ge A_{ILG0}\left(\frac{p_{IE1} p_{LC}^{\,2}}{z_{IE1} z_{CP}}\right) \approx f_{I0dB} \approx p_G$$

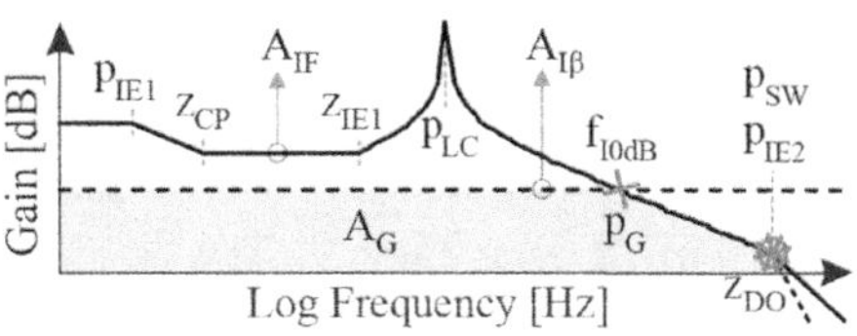

Example: Determine A_{IE0}, p_{IE2}, & A_{G0} so $f_{I0dB} = 10\% f_{SW}$ when $d_O = 33\%$, $p_{IE1} = z_{CP}$, $z_{IE1} = p_{LC}$ with other parameters from earlier examples.

Solution: $\beta_{IFB} = 1\ \Omega$, $f_{SW} = 1$ MHz from earlier example

$$L_{DO} = \frac{L_X}{D_O^2} = \frac{10\mu}{33\%^2} = 92\ \mu H \qquad R_{LO} = \frac{R_L}{D_O} = \frac{50m}{33\%} = 150\ m\Omega$$

$$A_{IL0} \approx \frac{V_E + V_D}{D_O^2 (R_{LO} + R_{LD})} = \frac{2+4}{33\%^2(150m+100)} = 550\ mA$$

$$p_{LC} = \frac{1}{2\pi\sqrt{L_{DO}C_O}} = \frac{1}{2\pi\sqrt{(92\mu)(5\mu)}} = 7.4\ kHz$$

$$p_{IE1} \equiv z_{CP} = \frac{1}{2\pi(R_C + R_{LD})C_O} = \frac{1}{2\pi(10m+100)(5\mu)} = 320\ Hz$$

$$f_{I0dB} \approx A_{ILG0}\left(\frac{p_{IE1}p_{LC}^2}{z_{IE1}z_{CP}}\right) = A_{ILG0}\left[\frac{(320)(7.4k)^2}{(7.4k)(320)}\right] \equiv \frac{f_{SW}}{10} = 100\ kHz$$

$\therefore$ $A_{ILG0} = 14$ V/V $= 23$ dB $\qquad p_{IE2} \geq f_{I0dB} = 100$ kHz

$$A_{ILG0} = A_{IE0}A_{PWM0}A_{IL0}\beta_{IFB} \approx A_{IE0}(2)(550m)(1) \equiv 14\ V/V$$

$\therefore$ $A_{IE0} = 13$ V/V $= 22$ dB

$$A_{G0} = (A_{IE0}A_{PWM0}A_{IL0}) \,||\, \frac{1}{\beta_{IFB}} \approx [(13)(2)(550m)] \,||\, 1 = 930\ mA/V < \frac{1}{\beta_{IFB}}$$

E. Discontinuous Conduction

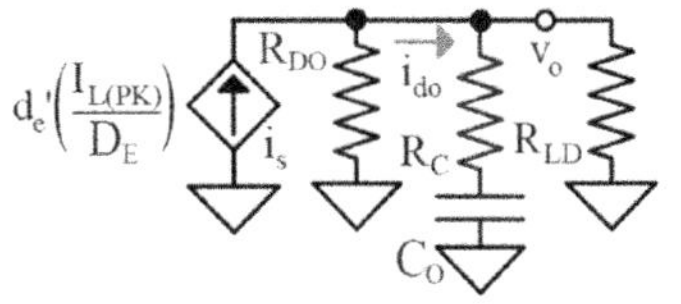

$$A_{IL} \equiv \frac{i_l}{d_e'} = \left(\frac{i_s}{d_e'}\right)\left(\frac{v_o}{i_s}\right)\left(\frac{i_{do}}{v_o}\right)\left(\frac{i_l}{i_{do}}\right) = \frac{i_{do}/D_O}{d_e'}$$

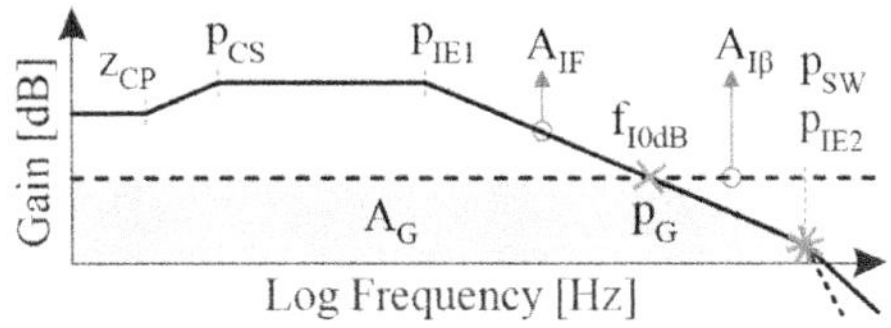

$$\approx \left(\frac{A_{LI}}{D_O}\right)\left[\frac{R_{DO} \,||\, (Z_C + R_C) \,||\, R_{LD}}{(Z_C + R_C) \,||\, R_{LD}}\right]$$

$$= \left(\frac{I_{L(PK)}}{D_E D_O}\right)\left(\frac{R_{DO} \,||\, R_{LD}}{R_{LD}}\right)\left[\frac{1 + s/2\pi z_{CP}}{(1 + s/2\pi p_{CS})(1 + s/2\pi p_{SW})}\right]$$

p_{IE1} reduces A_{ILG} to $f_{I0dB} = f_{IBW(CL)} = p_G \leq p_{IE2}, \frac{f_{SW}}{10}$

$$A_{ILG0}\Big|_{f_{I0dB} > p_{IE1} > p_{CS} > z_{CP}} \approx A_{ILG0}\left(\frac{p_{CS}p_{IE1}}{z_{CP}f_{I0dB}}\right) = 1 \quad \rightarrow \quad f_{I0dB} \approx A_{ILG0}p_{IE1}\left(\frac{p_{CS}}{z_{CP}}\right)$$

$\therefore$ A_G follows $A_{I\beta}$ up to p_G: $A_G \approx \dfrac{1/\beta_{IFB}}{(1 + s/2\pi p_G)(1 + s/2\pi p_{IE2})(1 + s/2\pi p_{SW})}$

6.6. Digital Control: A. Controller

Controller: Mixes, amplifies, stabilizes in digital domain

Voltage Controller:

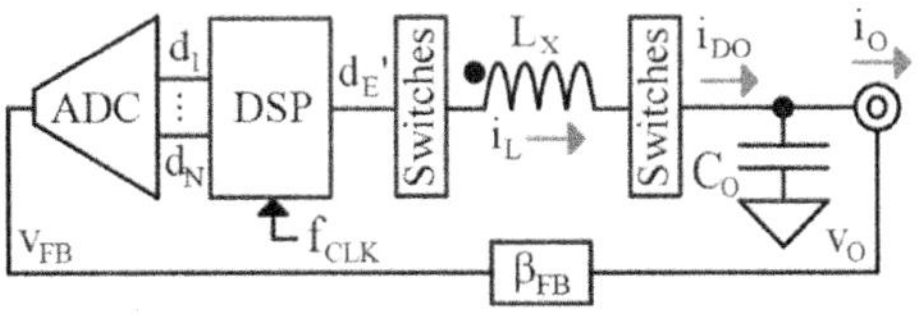

Current Controller:

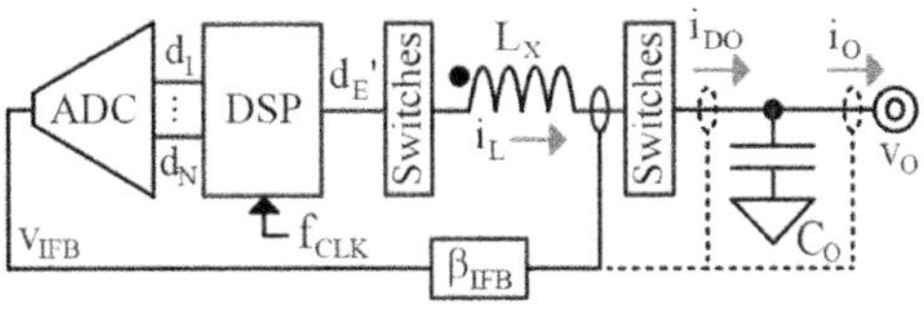

Response: $N_{LSB} \equiv$ LSB's needed to sweep d_E' from zero to 1 or 100%

$N_{CLK} \equiv$ Clock cycles needed to process a d_E' adjustment

$$A_{DIG} \equiv \frac{d_e'}{v_{fb}} = \frac{A_{DIG0}}{1 + s/2\pi p_{DIG}} \approx \frac{\Delta d_{E(MAX)}'/\Delta v_{FB}}{1 + s/2\pi f_{BW}} \approx \frac{1/(N_{LSB}v_{LSB})}{1 + s/2\pi (N_{CLK}t_{CLK})^{-1}}$$

Limit Cycling $\equiv d_E'$ alternates between nearest states

Tradeoffs: Programmable, flexible (share DSP), but large (A_{Si} for DSP), often slow

Chapter 7. Control Loops

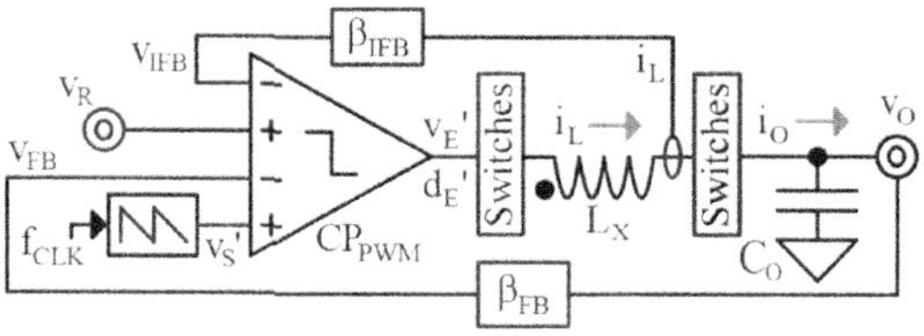

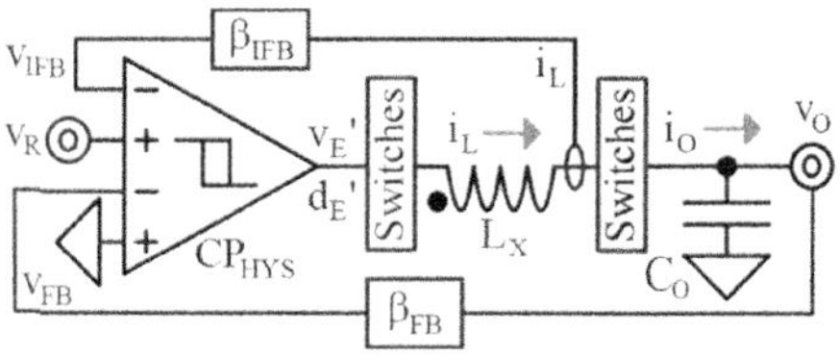

7.1. PWM (Average) Loop: A. Comparator

Function: Compare analog inputs | 1-bit analog–digital converter

"Trip" v_O when inputs crisscross | Polarity detector

Operation:

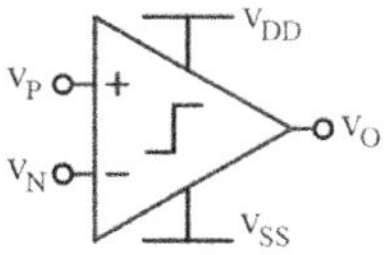

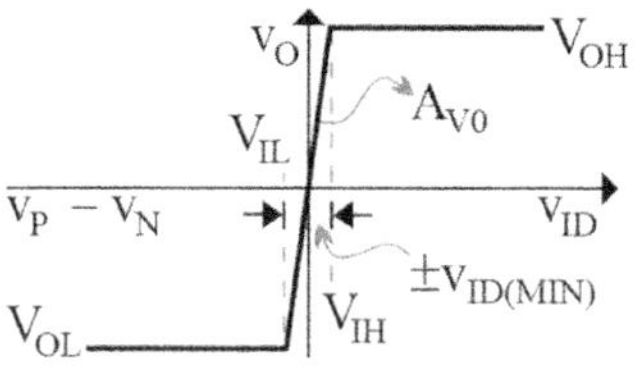

Static Parameters:

$V_{OH} \equiv \text{Output High} \approx v_{DD}$ $\qquad \pm v_{ID(MIN)} \equiv \text{Resolution}$

$V_{OL} \equiv \text{Output Low} \approx v_{SS}$ $\qquad \text{Gain} \equiv A_{V0} \equiv \frac{\Delta v_O}{\Delta v_{ID}} \approx \frac{V_{OH} - V_{OL}}{2 v_{ID(MIN)}}$

Propagation Delay:

$t_P \equiv$ Delay between

v_{ID}'s & v_O's halfway points

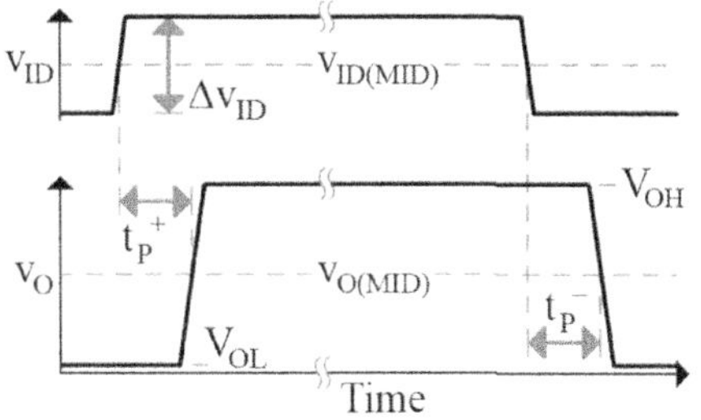

Dynamic Response:

Δv_{ID} = Small Signal → p_{BW} delays small signals

$$A_V = \frac{A_{V0}}{1+s/2\pi p_{BW}} = \frac{A_{V0}}{1+\tau_{BW}s} \qquad v_O = \overbrace{v_{ID}A_{V0}}^{v_{O(F)}}\left[1-\exp\left(\frac{-t}{\tau_{BW}}\right)\right]$$

Δv_O = Large Signal → $i_{C(MAX)}$ into C_X (slew rate) delays larger transitions

$$t_{P(SR)} = C_X\left(\frac{\Delta v_C}{i_{C(MAX)}}\right) \qquad \therefore \quad t_P \approx t_{P(BW)} + t_{P(SR)}$$

Bandwidth Delay: Defined with step response → Instant $\Delta v_{ID} \geq \Delta v_{ID(MIN)}$

$$\Delta v_{O(MID)} = \frac{\Delta v_{O(MAX)}}{2} = \frac{V_{OH}-V_{OL}}{2} = \Delta v_{ID}A_{V0}\left[1-\exp\left(\frac{-t_{P(BW)}}{\tau_{BW}}\right)\right]$$

Minimum drive: $\Delta v_{ID(MIN)} = 2v_{ID(MIN)} \equiv \dfrac{V_{OH}-V_{OL}}{A_{V0}}$ $\quad\therefore\quad \Delta v_{ID} = K_O \Delta v_{ID(MIN)}$

Overdrive factor: $K_O \equiv \dfrac{\Delta v_{ID}}{\Delta v_{ID(MIN)}} \geq 1$ $\qquad t_{P(BW)} = \tau_{BW}\ln\left(1-\dfrac{1}{2K_O}\right)^{-1} \leq 70\%\tau_{BW}$

Response: When $t_{P(BW)} >> t_{P(SR)}$, negative exponential → Initially fast, then slow

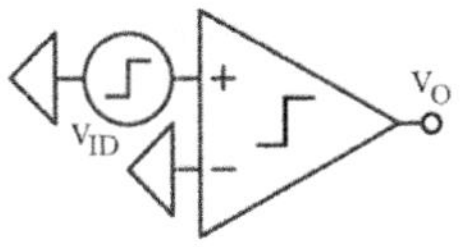

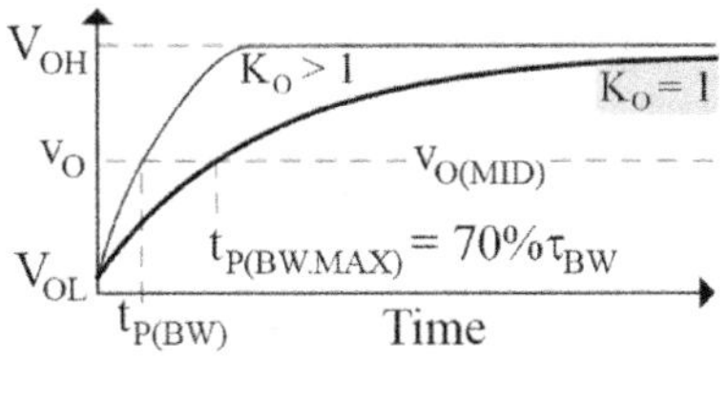

When $K_O > 1$

$\Delta v_{ID}A_{V0} > V_{OH} - V_{OL}$

$\therefore$ V_{OH} clamps v_O

B. Pulse-Width Modulator

Function: Set $d_O \equiv \frac{t_O}{t_{CLK}} \propto v_I$

Operation:

"Sawtooth" →

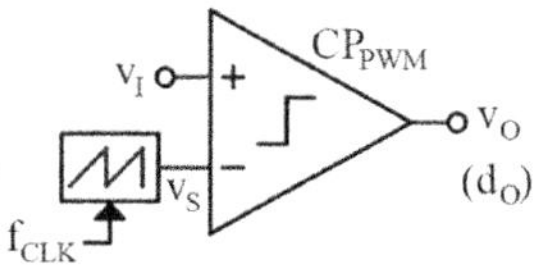

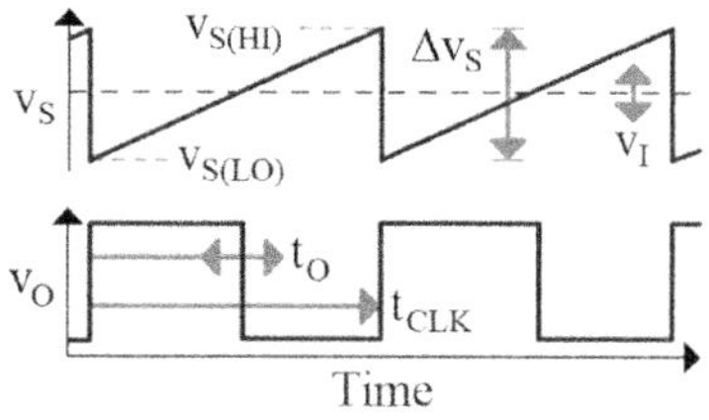

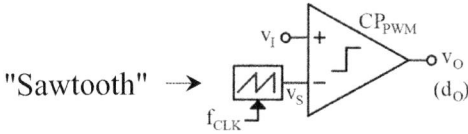

Static Translation:

$$A_{PWM0} \equiv \frac{\Delta d_O}{\Delta v_I} = \frac{1-0}{v_{S(HI)} - v_{S(LO)}} = \frac{1}{\Delta v_S}$$

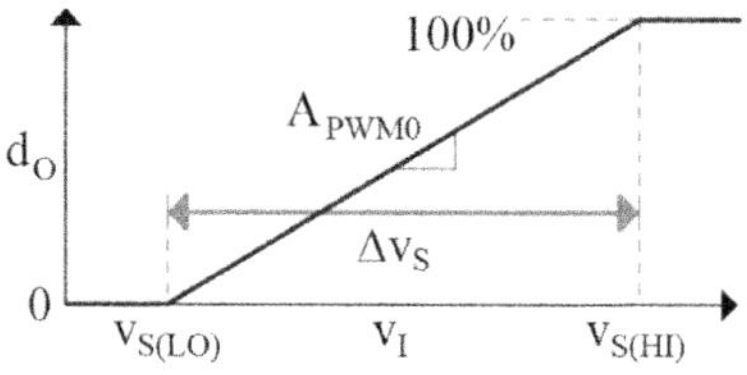

Dynamic Translation:

$$v_I = v_{S(LO)} + \left(t_P^+ + t_O - t_P^-\right)\left(\frac{dv_S}{dt}\right)$$

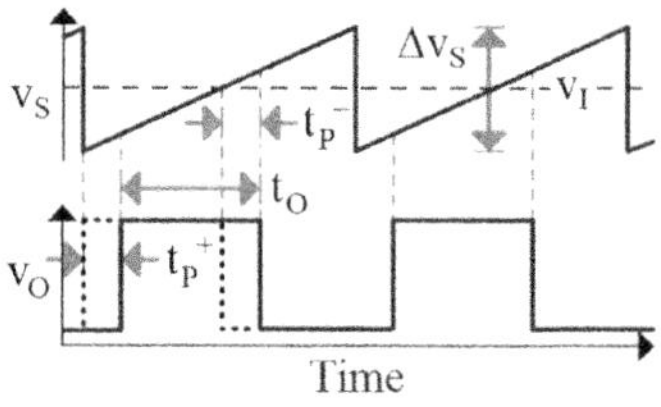

When $t_P^+ \approx t_P^-$:

$$\approx v_{S(LO)} + t_O\left(\frac{\Delta v_S}{t_{CLK}}\right) = v_{S(LO)} + d_O \Delta v_S$$

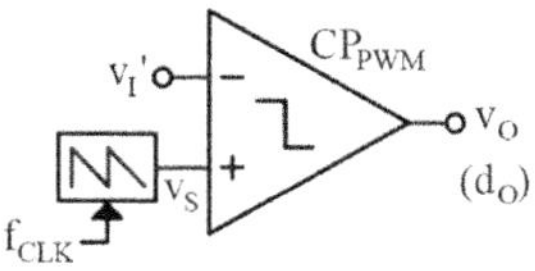

$$d_{O(MIN)} = \frac{t_{O(MIN)}}{t_{CLK}} = \frac{t_P^-}{t_{CLK}} \approx \frac{t_P}{t_{CLK}}$$

Inverting Translation:

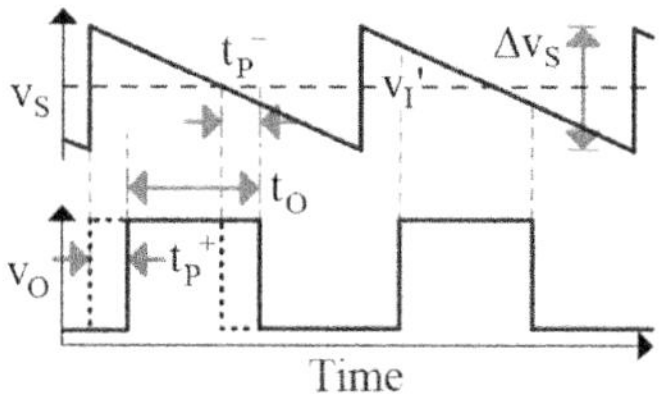

$$A_{PWM0} \equiv \frac{\Delta d_O}{\Delta v_I} = \frac{1-0}{v_{S(LO)} - v_{S(HI)}} = -\frac{1}{\Delta v_S}$$

$$v_I' \approx v_{S(HI)} + \left(t_P^+ + t_O - t_P^-\right)\left(\frac{dv_S}{dt}\right) \approx v_{S(HI)} - t_O\left(\frac{\Delta v_S}{t_{CLK}}\right) = v_{S(HI)} - d_O \Delta v_S$$

Example: Determine v_I & $d_{O(MIN)}$ when $v_{S(LO)} = 200$ mV, $v_{S(HI)} = 500$ mV, $d_O = 45\%$, $t_P = 100$ ns, $t_{CLK} = 1$ µs.

Solution: $\Delta v_S = v_{S(HI)} - v_{S(LO)} = 500m - 200m = 300$ mV

$v_I = v_{S(LO)} + d_O \Delta v_S = 200m + (45\%)(300m) = 335$ mV

$$d_{O(MIN)} = \frac{t_P}{t_{CLK}} = \frac{100n}{1\mu} = 10\%$$

SPICE Simulation:

* Non-Inverting PWM

v1v v1v 0 dc=1 → 1-V supply

vi vi 0 dc=335m

vs vs 0 dc=200m pulse 200m 500m 0 999n 1n 0n 1u

Sub-circuit ↑ Comparator ↑

xpwm vi vs vo v1v 0 cp

.lib lib.txt

.tran 5u

.end

* Plot v_I, v_S, v_O

C. PWM Loops

Voltage Loop:

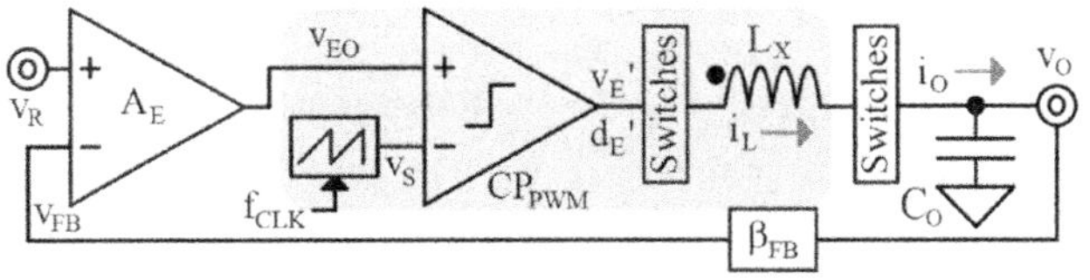

Current Loop:

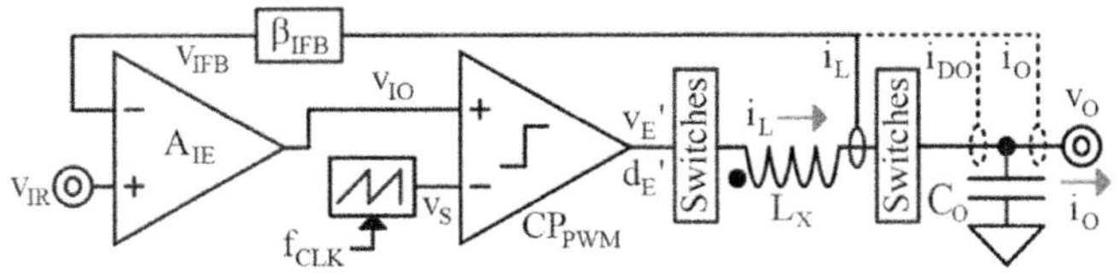

Current-Mode Voltage Loop: Embed i_L loop in v_O loop

$$A_G = \frac{i_{L(AVG)}}{v_{IR}}$$

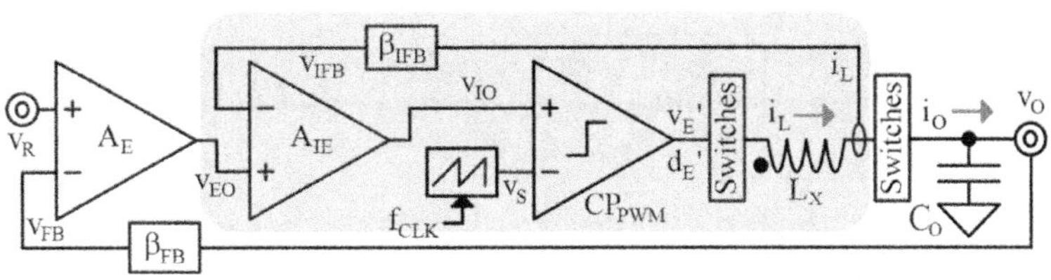

D. Gain Offsets

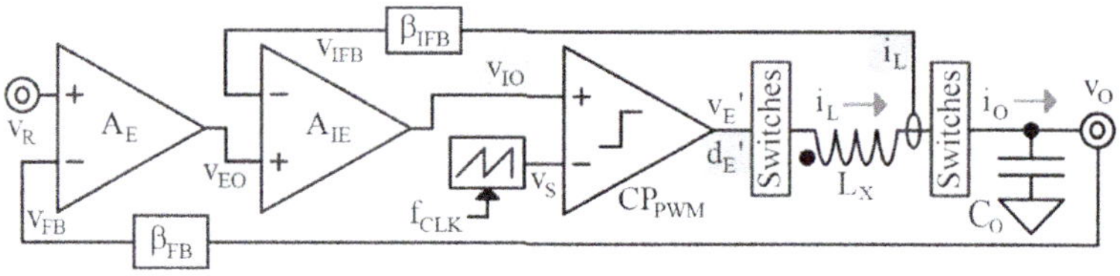

I-Loop Offset: v_{IO} needed for d_E' → $v_{IOS} \equiv v_{EO} - v_{IFB(AVG)} = \dfrac{v_{IO}}{A_{IE}} = \dfrac{v_{S(LO/HI)} \pm d_E'\Delta v_S}{A_{IE}}$

V-Loop Offset: v_{EO} needed for d_E' & $i_{L(AVG)}$ Loading Effect $\propto i_{L(AVG)} \propto i_O$

↓

$$v_{VOS} \equiv v_R - v_{FB(AVG)} = \frac{v_{EO}}{A_E} = \frac{v_{IOS} + v_{LD}}{A_E} = \frac{v_{IOS} + i_{L(AVG)}\beta_{IFB}}{A_E}$$

Compensation: Correct/center $v_{I/VOS}$ with β_{IFB} & β_{FB} translations

$$\beta_{IFB} \equiv \frac{\overline{v_{IFB(AVG)}}}{i_{L(AVG)/O}} = \frac{v_{EO} - \overline{v_{IOS}}}{i_{L(AVG)/O}} \qquad \beta_{FB} \equiv \frac{\overline{v_{FB(AVG)}}}{v_{O(AVG)}} = \frac{v_R - \overline{v_{VOS}}}{v_{O(AVG)}}$$

Example: Determine v_{IO}, v_{IOS}, v_{VOS}, $v_{FB(AVG)}$, β_{FB}, & v_O's error so v_O = 1.8 V

when v_R = 1.2 V, β_{IFB} = 1 Ω, $A_E = A_{IE}$ = 10 V/V, $i_{L(AVG)}$ = 100–500 mA,

$v_{S(LO)}$ = 200 mV, $v_{S(HI)}$ = 500 mV, d_E' = 45%, t_P = 100 ns, t_{CLK} = 1 μs.

Solution:

v_{IO} = 340 mV from earlier example → $v_{IOS} = \dfrac{v_{IO}}{A_{IE}} = \dfrac{340m}{10} = 34\ \text{mV}$

$$v_{VOS} = \frac{v_{IOS} + i_{L(AVG)}\beta_{IFB}}{A_E} = \frac{34m + i_{L(AVG)}(1)}{10} \approx 33 \pm 20\ \text{mV}$$

$$v_{FB(AVG)} = v_R - v_{VOS} = 1.2 - v_{VOS} = 1.17\ \text{V} \pm 20\ \text{mV}$$

$$\beta_{FB} \equiv \frac{\overline{v_{FB(AVG)}}}{v_{O(AVG)}} = \frac{1.17}{1.80} = 65\%$$ Loading Effect = ±1.7%

↑

$$v_{O(AVG)} = \frac{v_{FB(AVG)}}{\beta_{FB}} = \frac{v_{FB(AVG)}}{65\%} = 1.80\ \text{V} \pm 31\ \text{mV}$$

SPICE Tips

β_{FB}-translate v_O: efb vfb 0 vo 0 0.65 → V-controlled V source

Sense i_L: lx vswi vl 10u

vi vl vswo 0 → 0-V Current Sensor

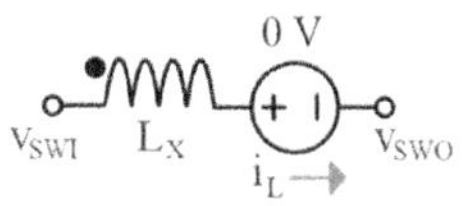

β_{IFB}-translate i_L: fbifb 0 vifb vi 1 → I-controlled I source

rbifb vifb 0 1 → β_{IFB}

Start Simulation: Slowly: vr vr 0 dc=1.2 pwl 0 0 1m 1.2

With R_{LD}: rld vo 0 18

Add i_O after: io vo 0 pwl 0 0 1.2m 0 1.3m 400m

Suppress Breakaway Excursions: Add capacitance to floating nodes

cfb vfb 0 1p

cifb vifb 0 1p

7.2. Hysteretic Loop: A. Hysteretic Comparator

Function: Split & shift rising & falling v_T's

→ Rise after v_{ID} rises over $v_{T(HI)}$ Fall after v_{ID} falls under $v_{T(LO)}$

Operation:

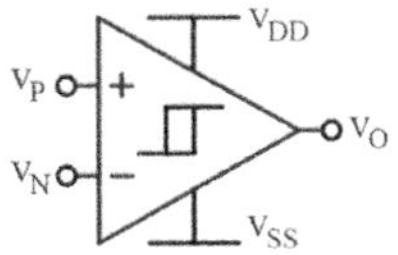

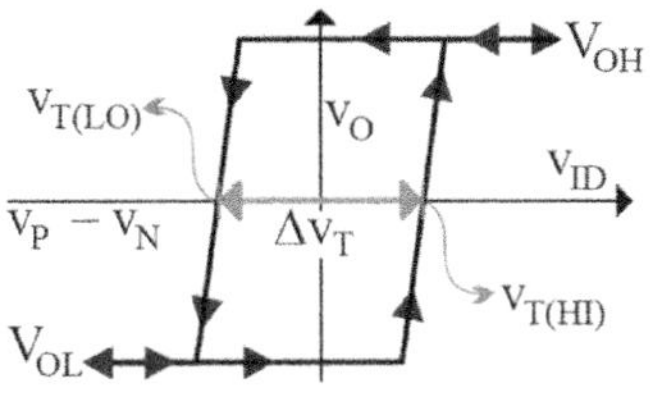

Δv_T = Hysteresis

Inverting v_O falls after $v_N - v_P$ rises over $-v_{T(LO)}$ → v_N's $v_{T(HI)} = -v_{T(LO)}$

Duality: v_O rises after $v_N - v_P$ falls under $-v_{T(HI)}$ → v_N's $v_{T(LO)} = -v_{T(HI)}$

B. Hysteretic Loops

Current Loop:

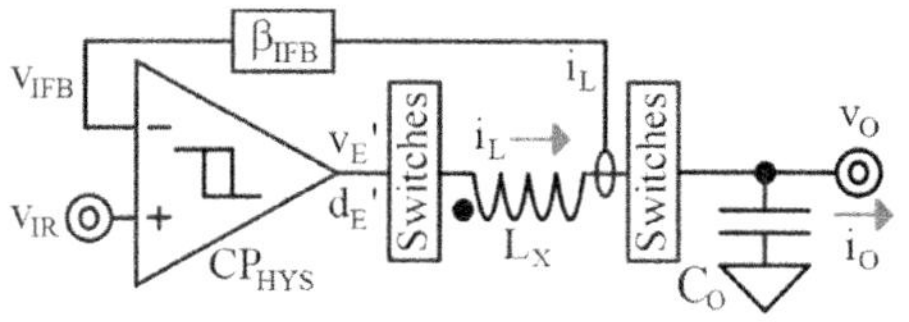

When v_{IFB} reaches $v_{IFB(LO)}$ → CP_{HYS} starts t_E

When v_{IFB} reaches $v_{IFB(HI)}$ → CP_{HYS} ends t_E

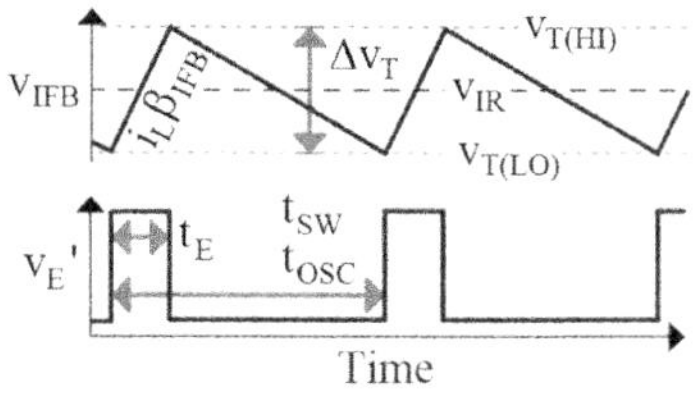

Relaxation Osc.: CP_{HYS} relaxes between transitions

Ring Oscillator: – FB, delay that inverts, & clamper → + FB A_{LG} at $f_{OSC} = 1$

↓ ↓ ↓

→ i_L reverses with CP_{HYS}, slews to opposite v_T, & v_T's set cycle–cycle gain = 1

Current-Mode Voltage Loop:

Embed i_L loop in v_O loop

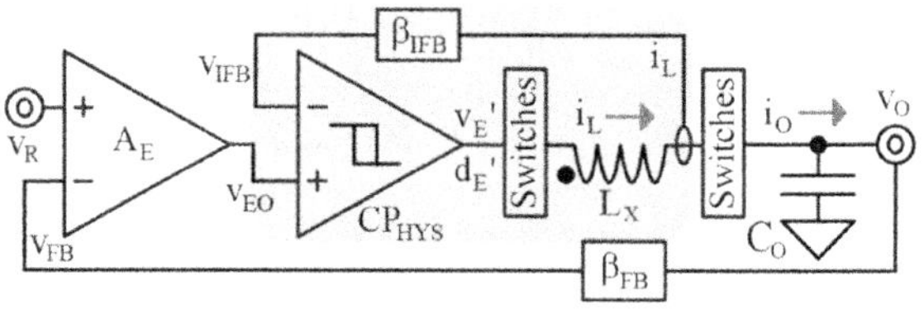

C. Projection Offsets

t_P shifts v_T's

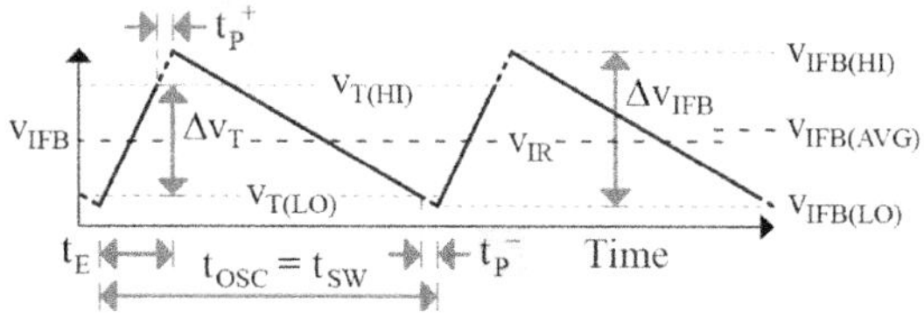

I Loop: $$v_{POS}^{\pm} = t_P^{\pm}\left(\frac{dv_{IFB}}{dt_{E/D}}\right) = t_P^{\pm}\left(\frac{di_L\beta_{IFB}}{dt_{E/D}}\right) = t_P^{\pm}\left(\frac{v_{E/D}}{L_X}\right)\beta_{IFB}$$

$$v_{IOS} \equiv v_{EO} - v_{IFB(AVG)} = -\Delta v_{IFB(AVG)} = -\left(\frac{v_{POS}^{+} - v_{POS}^{-}}{2}\right) \approx t_P\left(\frac{v_D - v_E}{2L_X}\right)\beta_{IFB}$$

When $t_P^{+} \approx t_P^{-}$

V Loop: $$v_{VOS} \equiv v_R - v_{FB(AVG)} = \frac{v_{EO}}{A_E} = \frac{v_{IOS} + v_{LD}}{A_E} = \frac{v_{IOS} + i_{L(AVG)}\beta_{IFB}}{A_E}$$

↑

Loading Effect ∝ $i_{L(AVG)}$

Compensation: Correct/center $v_{I/VOS}$ with β_{IFB} & β_{FB} translations

D. Oscillating Period

t_P expands $\Delta v_{IFB} = \Delta i_L \beta_{IFB} = \Delta v_T + v_{POS}^{+} + v_{POS}^{-} \approx \Delta v_T + t_P\left(\frac{v_E + v_D}{L_X}\right)\beta_{IFB}$

$$t_{SW} = \frac{t_E}{d_E} = \left(\frac{\Delta v_{IFB}}{d_E}\right)\left(\frac{dt_E}{di_L \beta_{IFB}}\right)$$

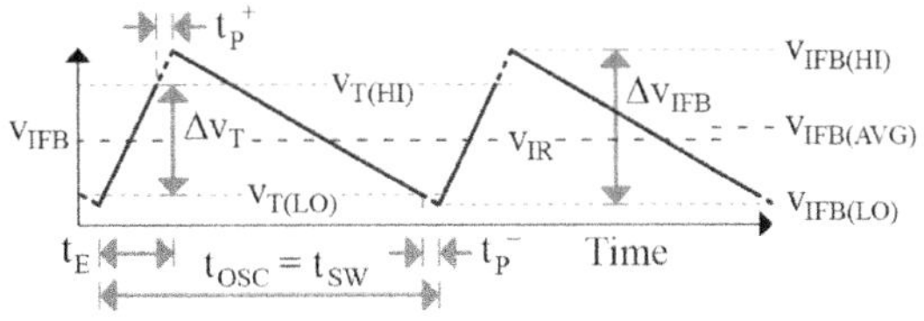

$$= \left(\frac{\Delta v_{IFB}}{d_E}\right)\left(\frac{L_X}{v_E}\right)\left(\frac{1}{\beta_{IFB}}\right)$$

Sensitive to:

$$= \left(\frac{\Delta v_{IFB}}{\beta_{IFB}}\right)\left(\frac{v_E + v_D}{v_D v_E}\right)L_X = \overbrace{\left(\frac{\Delta v_{IFB}}{\beta_{IFB}}\right)}^{\Delta i_L}\left(\frac{L_X}{v_E \parallel v_D}\right) = f(\ v_T\text{'s},\ t_P,\ v_E,\ v_D,\ L_X\ )$$

Note: If $t_{LC} << t_{SW}$ → Uncontrolled high-Δv_O oscillations at f_{LC}

If $t_{SW} << t_{LC}$ → Controlled low-Δv_O oscillations at f_{SW}

Example: Determine Δv_{IFB}, t_{SW}, v_{IOS}, v_{VOS}, $v_{FB(AVG)}$, β_{FB}, & v_O's error so

$v_O = 1.8$ V when $v_R = 1.2$ V, $A_E = 10$ V/V, $\beta_{IFB} = 1\ \Omega$, $\Delta v_T = 50$ mV,

$t_P = 100$ ns, $v_E = 2.2$ V, $v_D = 1.8$ V, $L_X = 10$ μH, $i_{L(AVG)} = 100$–500 mA.

Solution:

$$\Delta v_{IFB} = \Delta v_T + t_P\left(\frac{v_E + v_D}{L_X}\right)\beta_{IFB} = 50\text{m} + (100\text{n})\left(\frac{2.2 + 1.8}{10\mu}\right)(1) = 90\ \text{mV}$$

$$t_{SW} = \left(\frac{\Delta v_{IFB}}{\beta_{IFB}}\right)\left(\frac{L_X}{v_E \parallel v_D}\right) = \left(\frac{90\text{m}}{1}\right)\left(\frac{10\mu}{2.2 \parallel 1.8}\right) = 910\ \text{ns} \rightarrow f_{SW} = 1.1\ \text{MHz}$$

$$v_{IOS} = t_P\left(\frac{v_D - v_E}{2L_X}\right)\beta_{IFB} = -2\ \text{mV}$$

$$v_{VOS} = \frac{v_{IOS} + i_{L(AVG)}\beta_{IFB}}{A_E} = 30 \pm 20\ \text{mV}$$

$v_{FB(AVG)} = v_R - v_{VOS} = 1.2 - v_{VOS} = 1.17\ \text{V} \pm 20\ \text{mV}$

Same Loading Effect

↓

$$\beta_{FB} \equiv \frac{\overline{v_{FB(AVG)}}}{v_{O(AVG)}} = \frac{1.17}{1.80} = 65\%$$

$$v_{O(AVG)} = \frac{v_{FB(AVG)}}{\beta_{FB}} = \frac{v_{FB(AVG)}}{65\%} = 1.80\ \text{V} \pm 31\ \text{mV}$$

E. Response Time (Bandwidth)

Hysteretic Response:

$$\frac{\Delta v_{IR}}{\beta_{IFB}}$$

$$A_G \equiv \frac{i_{L(AVG)}}{v_{IR}} = \frac{A_{G0}}{1+s/2\pi p_{HYS}} = \frac{1/\beta_{IFB}}{1+\tau_{HYS}s} \qquad t_R = \frac{\Delta i_{L(AVG)}}{di_L/dt} = \Delta i_{L(AVG)}\left(\frac{L_X}{v_L}\right)$$

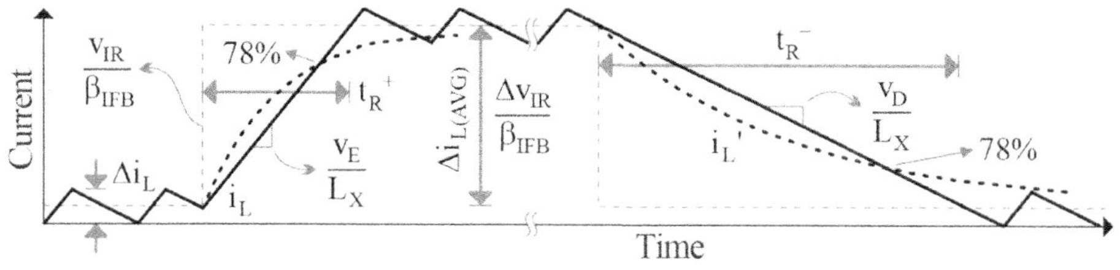

RC Model: $\tau_{HYS} \equiv 52\% t_R$ → $i_{L(78\%)} = i_{L(78\%)}'$ → Halves max. i_L error

$$p_{HYS} = \frac{1}{2\pi\tau_{HYS}} \approx \left(\frac{1}{2\pi}\right)\left(\frac{1.9}{t_R}\right) = \left(\frac{1.9}{2\pi}\right)\left(\frac{\beta_{IFB}}{\Delta v_{IR}}\right)\left(\frac{v_L}{L_X}\right) \propto \frac{v_L}{\Delta i_{L(AVG)}}$$

Fast with:
- High v_L
- Low $\Delta i_{L(AVG)}$

$$\frac{p_{HYS}}{f_{SW}} = p_{HYS}t_{SW} = \left(\frac{1.9}{2\pi}\right)\left(\frac{\Delta v_T'}{\Delta v_{IR}}\right)\left(\frac{v_{E/D}}{v_E \parallel v_D}\right)$$

→ $p_{HYS} \le f_{SW}$

→ p_{HYS} nears f_{SW} when Δv_{IR} = Low

Example: Determine t_{SW}, t_R, & p_{HYS} for the previous example when

Δv_T = 40–60 mV, t_P = 50–150 ns, L_X = 7–13 μH, $\Delta i_{L(AVG)}$ = ±400 mA.

Solution: v_E = 2.2 V, v_D = 1.8 V from previous example

$$\Delta v_{IFB} = \Delta v_T + t_P\left(\frac{v_E + v_D}{L_X}\right)\beta_{IFB} = 55\text{–}150 \text{ mV}$$

$$f_{SW} = \frac{1}{t_{SW}} = \left(\frac{\beta_{IFB}}{\Delta v_{IFB}}\right)\left(\frac{v_E \parallel v_D}{L_X}\right) = 510 \text{ kHz to } 2.6 \text{ MHz}$$ → Imprecise

$$t_R^+ = \Delta i_{L(AVG)}^+\left(\frac{L_X}{v_E}\right) = (+400\text{m})\left(\frac{L_X}{2.2}\right) = 1.3\text{–}2.4\ \mu\text{s}$$

$$t_R^- = \Delta i_{L(AVG)}^-\left(\frac{L_X}{v_D}\right) = (400\text{m})\left(\frac{L_X}{1.8}\right) = 1.6\text{–}2.9\ \mu\text{s}$$

$$p_{HYS} \approx \frac{1.9}{2\pi t_R} = 100\text{–}230 \text{ kHz}$$ → Within a decade or so of f_{SW}

7.3. Timed Peak/Valley Loops: A. SR Flip Flop

Function:	Remember previous digital state	]	Latch	State Machine
	Latch (clamp) output on-demand	]	Flip Flop	1-Bit Memory

Terminals: S ≡ Set R ≡ Reset Q ≡ Current State $\overline{Q}$ ≡ Opposite State

Operation:

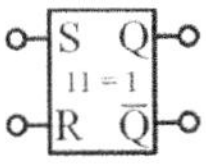

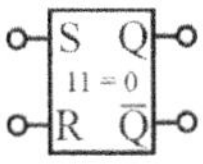

Hold when S = R = Low

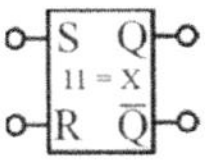

Reset Q when R = High

Set Q when S = High

(S)	R	Q
0	0	Q^{-1}
0	1	0
1	0	1
1	1	(1)

S	(R)	Q
0	0	Q^{-1}
0	1	0
1	0	1
1	1	(0)

S = R = High → Set-dominant sets   Reset-dominant resets

Purpose: Decouple set/reset commands ∴ Different loops/circuits control on/off

B. Pulse Generators

Set-Enabled:

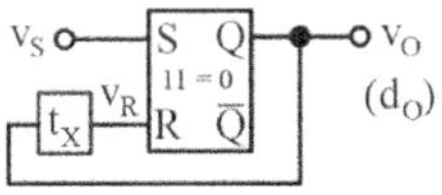

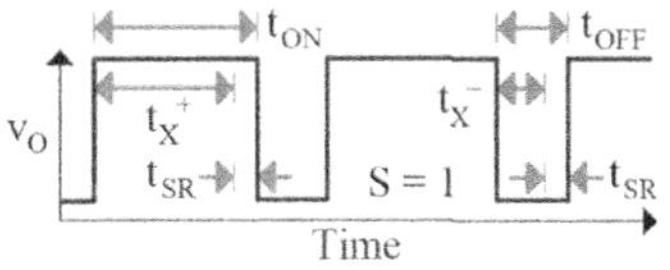

S = High & Latched – loop with delay → Ring oscillator $t_{ON} = t_X^+ + t_{SR}$

S = Low ∴ Can't set after Q resets → Oscillations stop $t_{OFF} = t_X^- + t_{SR}$

Reset-Enabled:

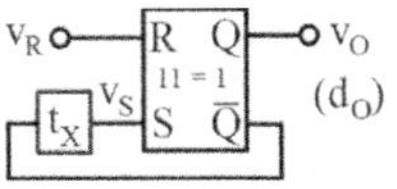

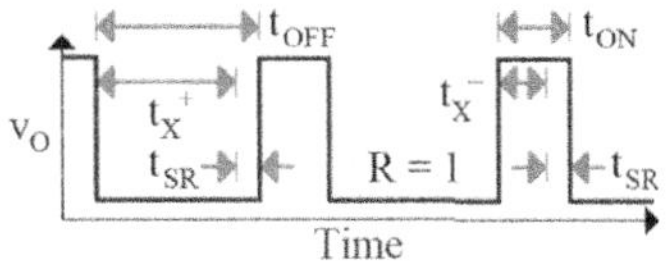

R = High & Latched – loop with delay → Ring oscillator $t_{OFF} = t_X^+ + t_{SR}$

R = Low ∴ Can't reset after Q sets → Oscillations stop $t_{ON} = t_X^- + t_{SR}$

Duty-Cycle: $d_{ON/OFF} = \frac{t_{ON/OFF}}{t_{OSC}} = \frac{t_{ON/OFF}}{t_{ON} + t_{OFF}} = \frac{t_X^+ + t_{SR}}{t_X^+ + t_X^- + 2t_{SR}}$

C. Constant On-Time Valley Loop

Valley Control { CP_T sets $v_{IFB(LO)}$; t_{ON} sets $v_{IFB(HI)}$ }

Current Loop: Decouple hys. v_T's

v_{IFB} falls below v_{IR} : v_{IO} starts $t_E = t_{ON}$: v_{IFB} rises, v_{IO} falls, Oscillator stops → One-Shot Pulse

t_E ends after t_X' : v_{IFB} falls w/ i_L : Sequence repeats

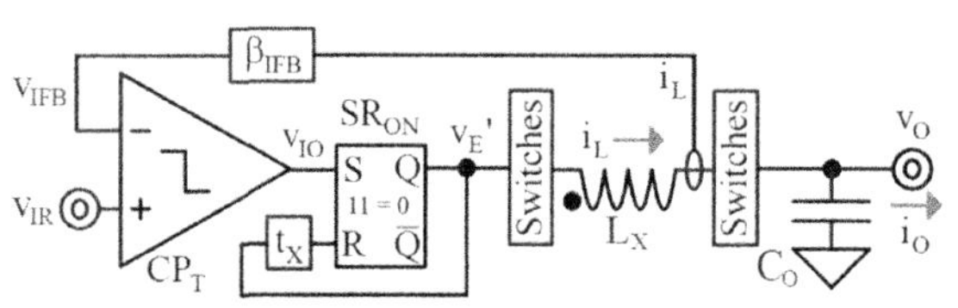

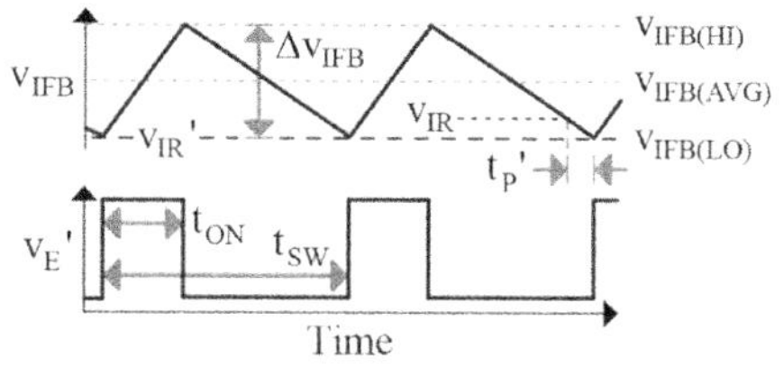

$$\Delta v_{IFB} = \left(\frac{di_L}{dt_E}\right)\beta_{IFB} t_E = \left(\frac{v_E}{L_X}\right)\beta_{IFB} t_{ON}$$

$$t_{SW} = \frac{t_E}{d_E} = t_{ON}\left(\frac{v_E + v_D}{v_D}\right)$$

$$v_{IOS} \equiv v_{IR} - v_{IFB(AVG)}$$

$$= -\frac{\Delta v_{IFB}}{2} + v_{POS}^{-} = -\frac{\Delta v_{IFB}}{2} + \left(\frac{v_D}{L_X}\right)\beta_{IFB} t_P'$$

$$v_{VOS} \equiv v_R - v_{FB(AVG)}$$

$$= \frac{v_{EO}}{A_E} = \frac{v_{IOS} + v_{LD}}{A_E}$$

Response Time:

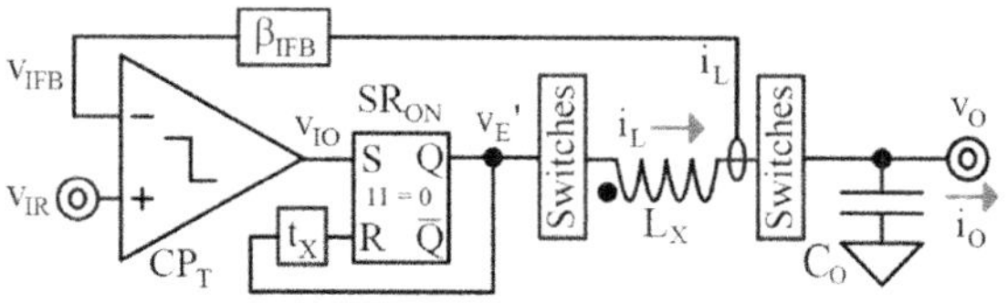

t_R^+ when v_{IR} steps up

→ v_{IO} rises & stays high

v_E' pulses — Requirement: i_L Rises ↓

t_{OFF} interrupts i_L's climb → Short t_{OFF}

t_R^- when v_{IR} steps down

→ v_{IO} falls & stays low

v_E' stops pulsing

i_L falls without interruptions

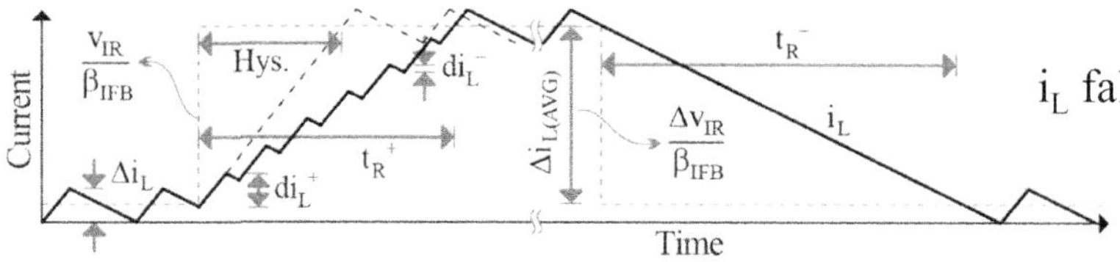

∴ Rises with short interruptions; Falls without interruptions } Fast when $t_{OFF} \ll t_{ON}$ — $f(t_{ON}, d_E)$; Slower than hyst., more predictable f_{SW}

D. Constant Off-Time Peak Loop

Current Loop: Decouple hys. v_T's

CP_T sets $v_{IFB(HI)}$ | Peak Control
t_{OFF} sets $v_{IFB(LO)}$

v_{IFB} rises over v_{IR} : v_{IO} starts $t_D = t_{OFF}$: v_{IFB} falls, v_{IO} falls, & Oscillator stops → One-Shot Pulse

t_D ends after t_X : v_{IFB} rises w/ i_L : Sequence repeats

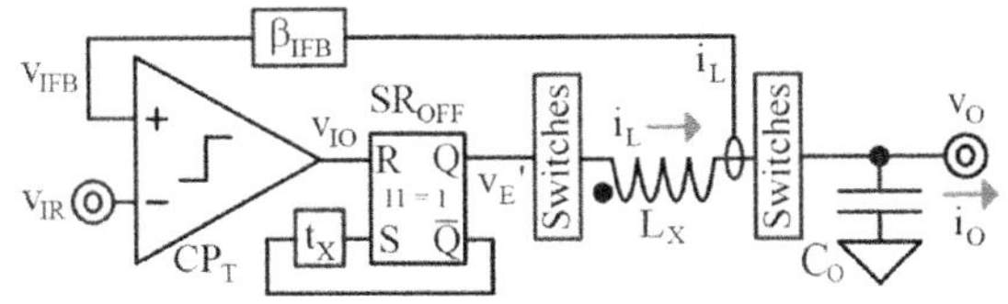

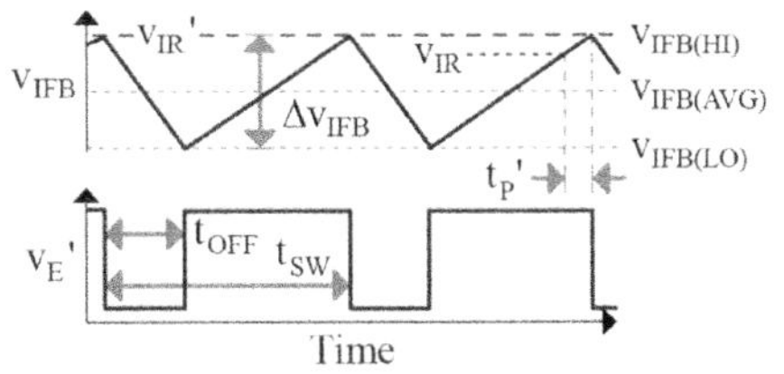

$$\Delta v_{IFB} = \left(\frac{di_L}{dt_D}\right)\beta_{IFB} t_D = \left(\frac{v_D}{L_X}\right)\beta_{IFB} t_{OFF}$$

$$t_{SW} = \frac{t_D}{d_D} = t_{OFF}\left(\frac{v_E + v_D}{v_E}\right)$$

$$v_{IOS} \equiv v_{IR} - v_{IFB(AVG)}$$

$$= \frac{\Delta v_{IFB}}{2} - v_{IOS}^{+} = \left[\left(\frac{v_D}{L_X}\right)\left(\frac{t_{OFF}}{2}\right) - \left(\frac{v_E}{L_X}\right)t_P'\right]\beta_{IFB}$$

$$v_{VOS} \equiv v_R - v_{FB(AVG)}$$

$$= \frac{v_{EO}}{A_E} = \frac{v_{IOS} + v_{LD}}{A_E}$$

Response Time:

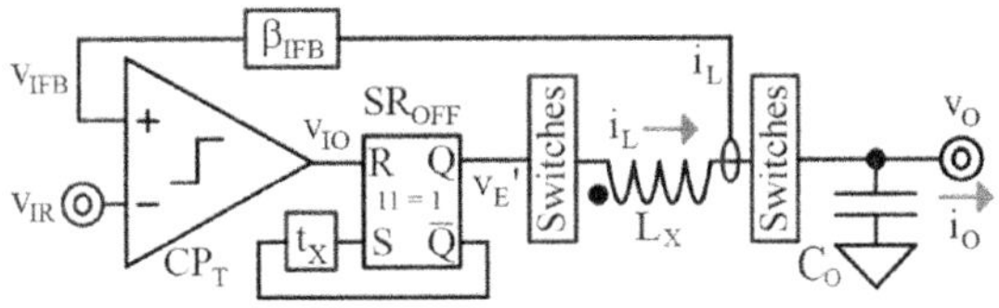

t_R^+ when v_{IR} steps up

→ v_{IO} falls & stays low

v_E' stops pulsing low (stays high)

i_L rises without interruptions

t_R^- when v_{IR} steps down

→ v_{IO} rises & stays high

v_E' pulses

t_{ON} interrupts i_L's fall

→ Short t_{ON}

↑

Requirement: i_L falls

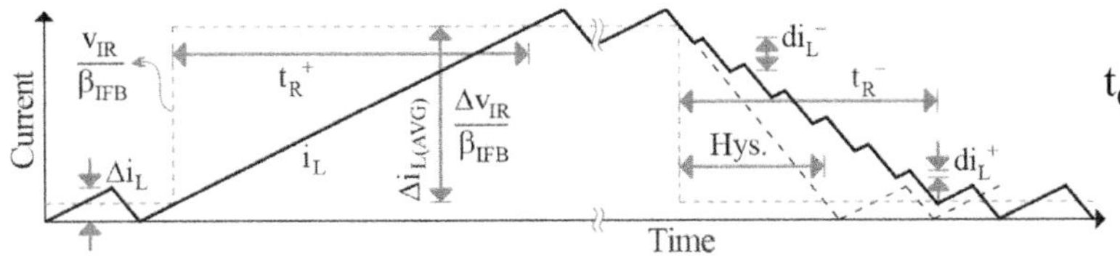

∴ Rises without interruptions

Falls with short interruptions → Fast when $t_{ON} << t_{OFF}$

E. Constant-Period Loops

Peak-Current Loop:

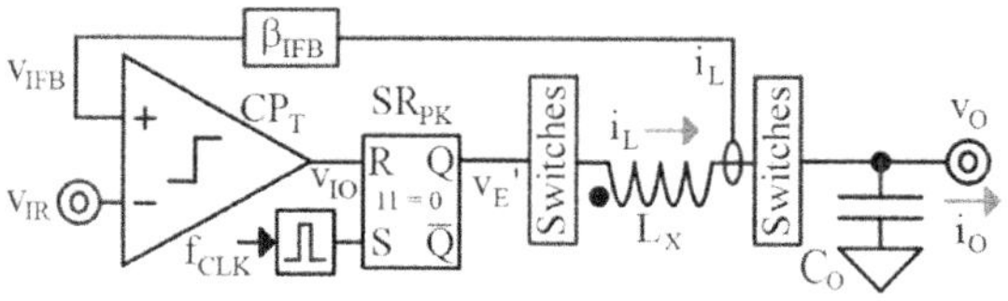

CP_T sets $v_{IFB(HI)}$ t_{CLK} sets $v_{IFB(LO)}$

→ v_{IFB} rises over v_{IR} : v_{IO} starts t_D : v_{IFB} falls : v_{IO} falls : t_{CLK} ends t_D : Repeats

Sudden v_{IR} rise/fall → Low v_{IO} ∴ Can't reset High v_{IO} ∴ Can't set

→ i_L rises/falls without interruptions → Hysteretic t_R (fast)

Valley-Current Loop:

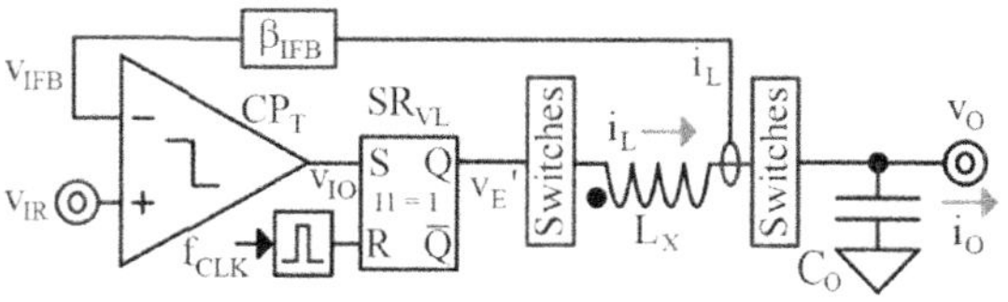

CP_T sets $v_{IFB(LO)}$ t_{CLK} sets $v_{IFB(HI)}$

→ v_{IFB} falls below v_{IR} : v_{IO} starts t_E : v_{IFB} rises : v_{IO} falls : t_{CLK} ends t_E : Repeats

Sudden v_{IR} rise/fall → High v_{IO} ∴ Can't reset Low v_{IO} ∴ Can't set

→ i_L rises/falls without interruptions → Hysteretic t_R (fast)

Sub-Harmonic Oscillation: Peak-Current Example

v_{IN} noise alters i_L's trajectory

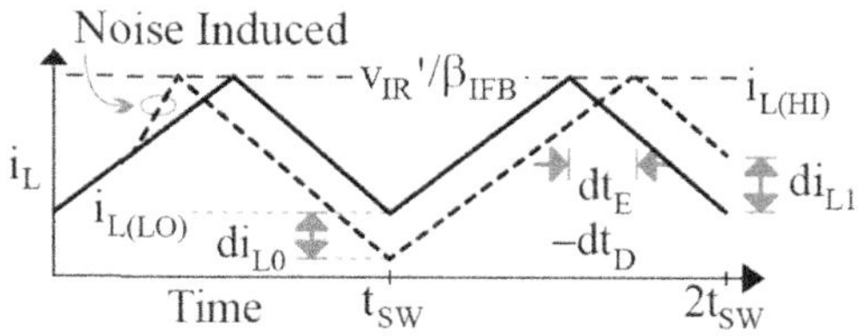

→ Induce initial imbalance di_{L0}

Imbalance inverts & repeats in di_{L1}

$$di_{L1} = di_{L0}\left(\frac{dt_E}{di_L}\right)\left(\frac{dt_D}{dt_E}\right)\left(\frac{di_L}{dt_D}\right)$$

∴ $d_E < 50\%$ → $A_X < 1$ ∴ di_L shrinks

$$= di_{L0}\left(\frac{L_X}{v_E}\right)(-1)\left(\frac{v_D}{L_X}\right)$$

$d_E = 50\%$ → $A_X = 1$ ∴ di_L repeats

$$= -di_{L0}\left(\frac{d_E}{d_D}\right) = -di_{L0}\overbrace{\left(\frac{d_E}{1-d_E}\right)}^{A_X}$$

$d_E > 50\%$ → $A_X > 1$ ∴ di_L grows

di_L repeats every other cycle → $f_{OSC} = 0.5f_{SW}$ ∴ Sub-harmonic oscillation

Slope Compensation:

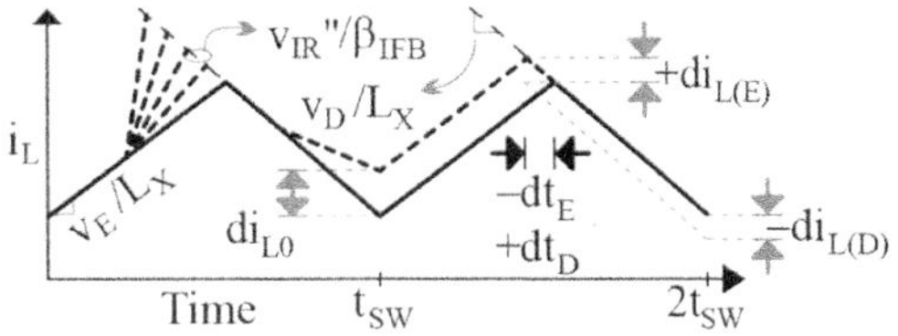

Slope v_{IR} to $\frac{di_L}{dt_D}$ → Subtract + slope from v_{IR}

$v_{IDI} = v_{IR} - v_{IFB}$ → Add + slope to v_{IFB}

$$di_{L1} = di_{L(E)} + di_{L(D)} = dt_E\left(\frac{di_L^*}{dt_E}\right) + dt_D\left(\frac{di_L}{dt_D}\right) = dt_E\left(\frac{di_L^*}{dt_E} - \frac{di_L}{dt_D}\right) = 0 \text{ when } \frac{di_L^*}{dt_E} \equiv \frac{di_L}{dt_D}$$

$$\therefore \quad dt_E = di_{L0}\left(\frac{dt_E}{di_L}\right) = di_{L0}\left(\frac{v_E}{L_X} + \frac{di_L^*}{dt_E}\right)^{-1}$$

↳ Add + Slope

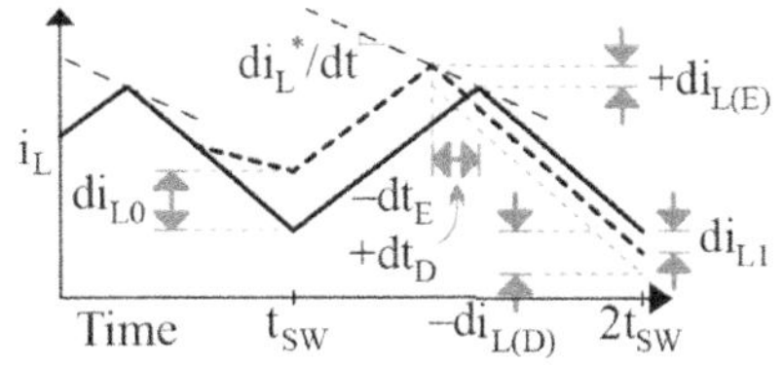

$$di_{L1} = di_{L0}\left(\frac{0.5v_D - v_D}{v_E + 0.5v_D}\right) = \frac{-di_{L0}}{2(d_D/d_E)+1}$$

Half Slope: $\frac{di_L^*}{dt_E} \equiv \left(\frac{1}{2}\right)\left(\frac{di_L}{dt_D}\right) = \frac{0.5v_D}{L_X}$ Denominator > 1 $\therefore$ di_{L1} shrinks

Slope-Compensated Current Loops:

Subtract + sawtooth from v_{IR}

Or add – sawtooth to v_{IR}

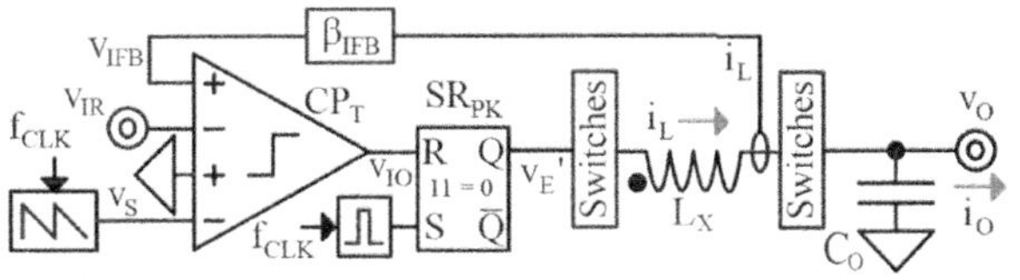

Peak Example

$$v_{IOS} \equiv v_{IR} - v_{IFB(AVG)} = v_{IOS(ON/OFF)} \pm v_{SOS} = \left(\mp\frac{\Delta v_{IFB}}{2} \pm v_{IOS}^{\mp}\right) \pm \left(v_{S(LO/HI)} \pm d_E'\Delta v_S\right)$$

Example: Determine $v_{S(HI)}$, Δv_{IFB}, v_{IOS}, v_{VOS}, & $v_{FB(AVG)}$ when $v_R = 1.2$ V,

$A_E = 10$ V/V, $\beta_{IFB} = 1\ \Omega$, $t_{CLK} = 1$ µs, $t_P = 100$ ns, $t_{SR} = 10$ ns, v_S starts from

$v_{S(LO)} = 200$ mV, $v_E = 2.2$ V, $v_D = 1.8$ V, $L_X = 10$ µH, $i_{L(AVG)} = 100$–500 mA.

Solution:

$$\frac{dv_{IFB}^*}{dt} = \frac{\Delta v_S}{t_{CLK}} = \frac{\Delta v_S}{1\mu} \equiv \frac{0.5v_D}{L_X} = \frac{0.5(1.8)}{10\mu} \quad \therefore \ \Delta v_S = 90 \text{ mV}$$

$$v_{S(HI)} = v_{S(LO)} + \Delta v_S = 200m + 90m = 290 \text{ mV}$$

$$d_E = \frac{v_D}{v_E + v_D} = \frac{1.8}{2.2+1.8} = 45\% \qquad t_E = d_E t_{CLK} = (45\%)(1\mu) = 450\ \text{ns}$$

$$\Delta v_{IFB} = t_E\left(\frac{v_E}{L_X}\right)\beta_{IFB} = (450n)\left(\frac{2.2}{10\mu}\right)(1) = 99\ \text{mV}$$

$$t_P' = t_P + t_{SR} = 100n + 10n = 110\ \text{ns} \qquad v_{SOS} = v_{S(LO)} + d_E \Delta v_S = 240\ \text{mV}$$

$$v_{IOS} = \frac{\Delta v_{IFB}}{2} - t_P'\left(\frac{v_E}{L_X}\right)\beta_{IFB} + v_{SOS} = \frac{99m}{2} - (110n)\left(\frac{2.2}{10\mu}\right)(1) + 240m = 265\ \text{mV}$$

$$v_{VOS} = \frac{v_{IOS} + i_{L(AVG)}\beta_{IFB}}{A_E} = \frac{265m + i_{L(AVG)}(1)}{10} = 56 \pm 20\ \text{mV}$$

$$v_{FB(AVG)} = v_R - v_{VOS} = 1.2 - v_{VOS} = 1.14\ \text{V} \pm 20\ \text{mV}$$

If contracted → Higher loading effect   If v_{LD}-compensated → No loading effect

7.4. Summing Contractions: A. Summing Comparator

Function: Sum v_P's & v_N's

Trip when sum crosses zero

→ Detect polarity of sum

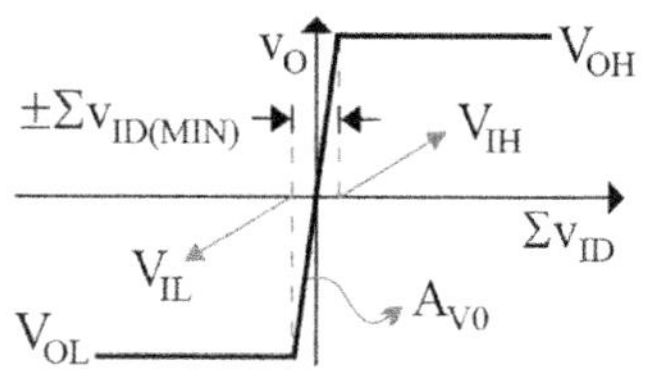

Differential Sum $\Sigma v_{ID} = (v_{P1} + v_{P2}) - (v_{N1} + v_{N2})$

$= (v_{P1} + v_{ID2}) - v_{N1}$ → Add v_{ID2} to v_{P1}

$= v_{ID1} + v_{ID2}$ → Analog Summer

Equivalence:

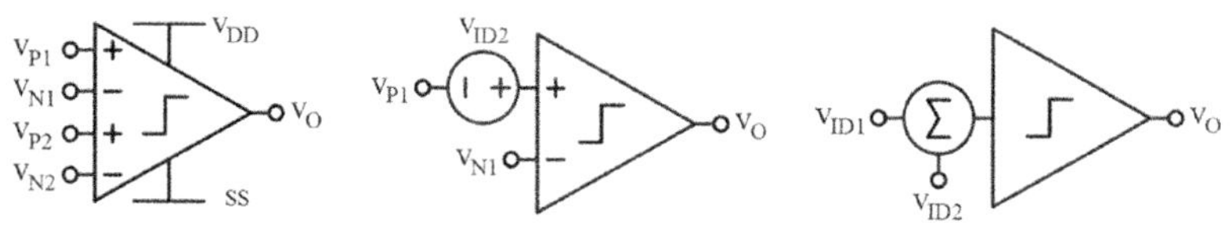

B. PWM Contractions

Condition: When gain & stabilizer in $A_{E/IE}$ are not necessary → $A_{E/IE} = 1$ V/V

Voltage Loop:

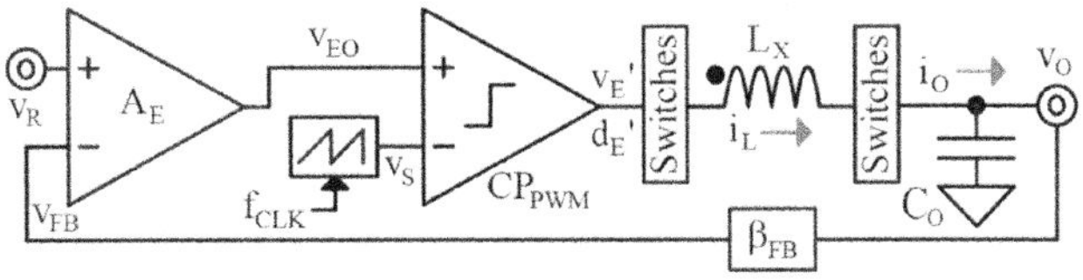

$$\Sigma v_{ID} = v_{EO} - v_S$$

$$= (v_R - v_{FB}) A_E \big|_{A_E \equiv 1} - v_S$$

$$= v_R - v_{FB} + 0 - v_S$$

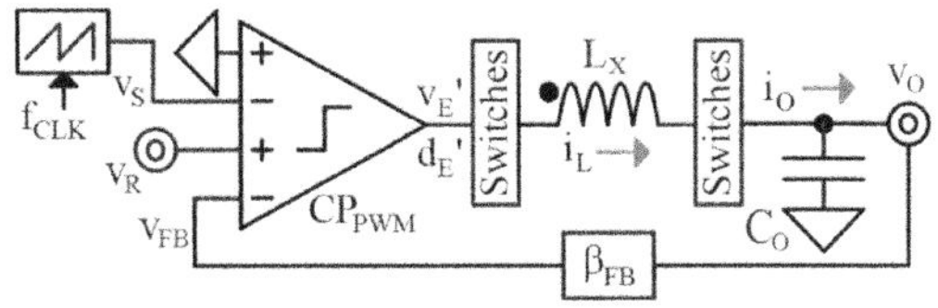

Current Loop:

$$\Sigma v_{ID} = v_{IO} - v_S$$

$$= (v_{IR} - v_{IFB}) A_{IE} \big|_{A_{IE} \equiv 1} - v_S$$

$$= v_{IR} - v_{IFB} + 0 - v_S$$

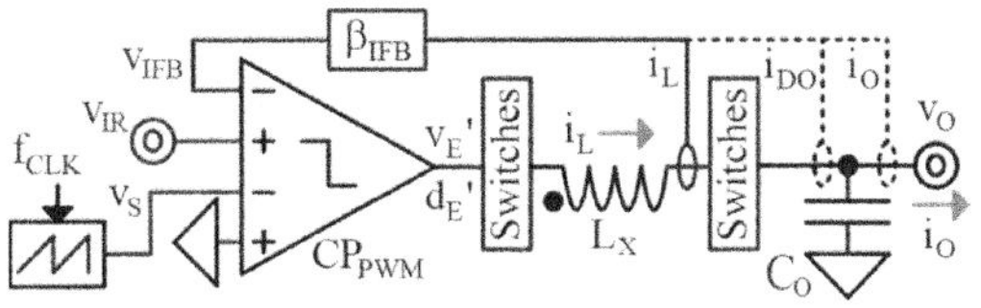

i_L-Mode v_O Loop: Embed i_L loop in v_O loop

$$A_G = \frac{i_{L(AVG)}}{v_{EO}} \qquad \therefore$$

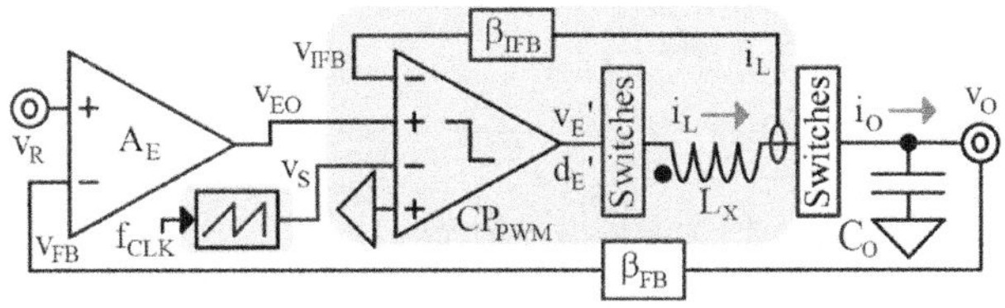

Double Contraction:

$$\Sigma v_{ID} = v_{EO} - v_{IFB} - v_S$$

$$= (v_R - v_{FB}) A_E \big|_{A_E \equiv 1} - v_{IFB} - v_S$$

$$= v_R - v_{FB} + 0 - v_{IFB} + 0 - v_S$$

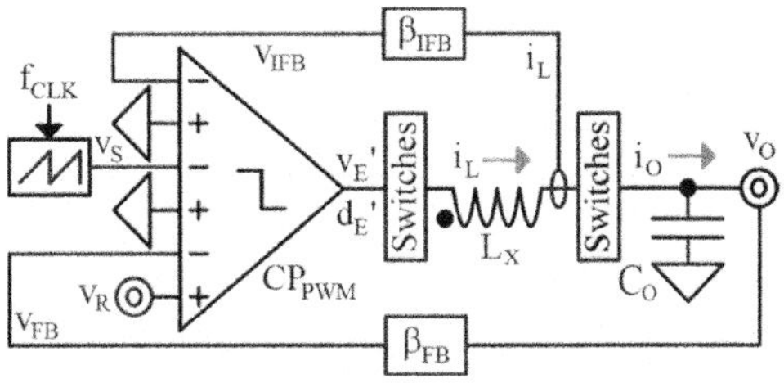

Compact Double Contraction:

$$= v_R - v_{FB} + (-v_S) - v_{IFB}$$

$$\downarrow$$

Invert v_S

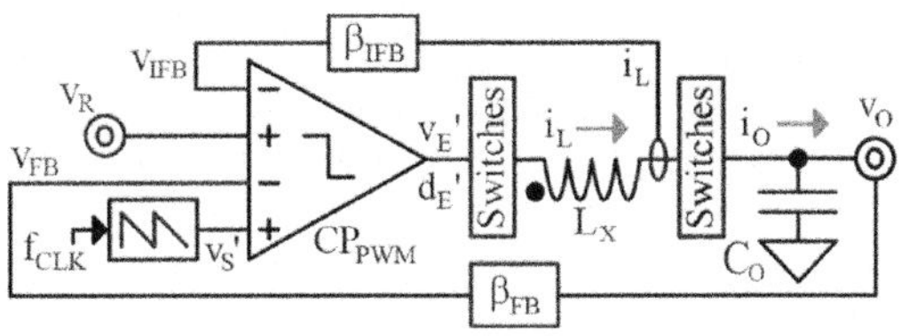

C. Gain Offsets

I-Loop Offset: $v_{IOS} = v_{IO} = v_{S(LO/HI)} \pm d_E'\Delta v_S$

V-Loop Offset: $v_{VOS} = v_{LD} \pm v_{IOS} = i_{L(AVG)}\beta_{IFB} \pm v_{IOS}$

} Unsuppressed in contractions

Example: Determine v_{IOS}, v_{VOS}, $v_{FB(AVG)}$, & v_O's error so $v_O = 1.8$ V when

$v_R = 1.2$ V, $d_E' = 45\%$, $v_S = 200$–500 mV, $\beta_{IFB} = 1\ \Omega$, $i_{L(AVG)} = 100$–500 mA.

Solution: $\Delta v_S = v_{S(HI)} - v_{S(LO)} = 500\text{m} - 200\text{m} = 300\text{ mV}$

$$v_{IOS} = v_{S(LO)} + d_E'\Delta v_S = 200\text{m} + (45\%)(300\text{m}) = 335\text{ mV}$$

$$v_{VOS} \approx v_{IOS} + i_{L(AVG)}\beta_{IFB} = 335\text{m} + i_{L(AVG)}(1) = 635 \pm 200\text{ mV}$$

$$v_{FB(AVG)} = v_R - v_{VOS} = 1.2 - v_{VOS} \approx 565 \pm 200\text{ mV}$$

$$\beta_{FB} \equiv \frac{\overline{v_{FB(AVG)}}}{v_{O(AVG)}} = \frac{565\text{m}}{1.80} = 31.4\% \quad \rightarrow \quad v_{O(AVG)} = \frac{v_{FB(AVG)}}{\beta_{FB}} = 1.80\text{ V} \underbrace{\pm\, 640\text{ mV}}_{\text{Inaccurate}}$$

D. Hysteretic Contraction

Condition: When gain & stabilizer in A_E are not necessary → $A_E = 1$ V/V

Current-Mode Voltage Loop:

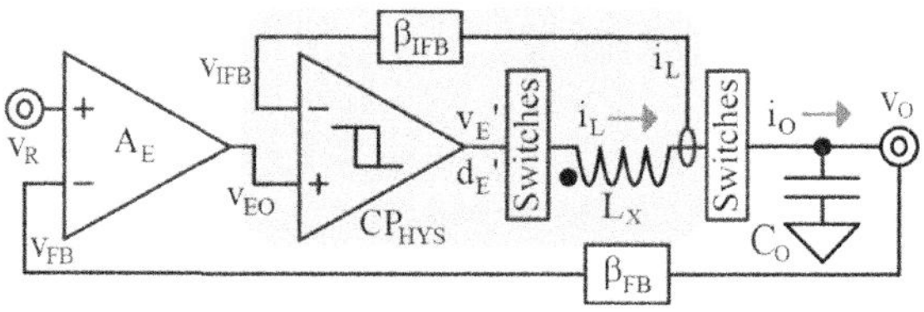

$$\Sigma v_{ID} = v_{EO} - v_{IFB}$$

$$= (v_R - v_{FB})A_E\big|_{A_E \equiv 1} - v_{IFB}$$

$$= v_R - v_{FB} + 0 - v_{IFB}$$

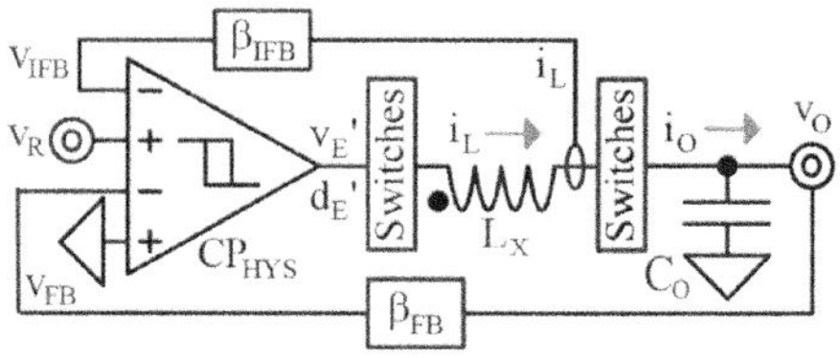

Voltage-Loop Offset: $A_E = A_{IE} = 1$ V/V

$$v_{VOS} \equiv v_R - v_{FB(AVG)} = v_{IOS} + v_{LD} \rightarrow \text{Unsuppressed}$$

E. Timed Contractions

Constant-t_{OFF} Peak i_L-Mode v_O Loop:

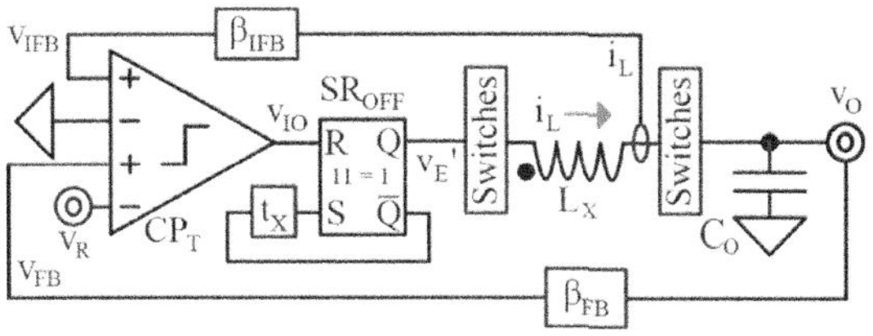

$$\Sigma v_{ID} = v_{IFB} - v_{EO}$$

$$= v_{IFB} - (v_R - v_{FB}) A_E \big|_{A_E \equiv 1}$$

$$= v_{IFB} - 0 + v_{FB} - v_R$$

Slope-Compensated

Constant-t_{SW} Peak i_L Loop:

Add – sawtooth to v_{IR}

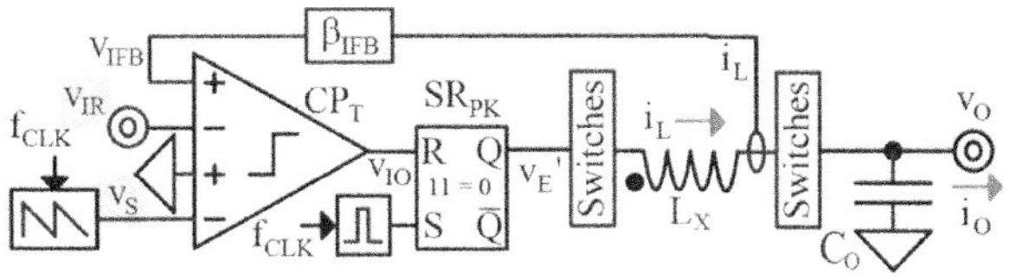

i_L-Mode v_O Loop:

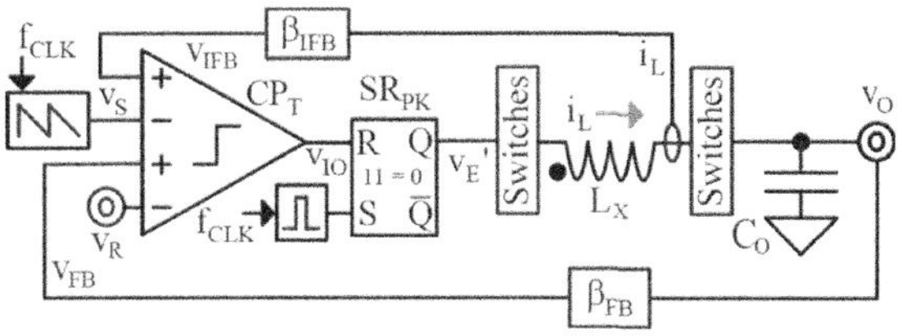

$$\Sigma v_{ID} = v_{IFB} - v_{EO} - (-v_S)$$

$$= v_{IFB} - (v_R - v_{FB}) A_E \big|_{A_E \equiv 1} + v_S$$

$$= v_{IFB} - (-v_S) + v_{FB} - v_R$$

F. Load Compensation: i. Principle

Problem: β_{FB} can center, but not cancel v_{LD} → Error = $\pm 0.5 v_{LD}$

Note: $v_{IFB} = i_L \beta_{IFB}$ carries $v_{LD} = i_{L(AVG)} \beta_{IFB}$

Fix: Average v_{IFB} & add $v_{IFB(AVG)}$ to v_R → $v_{LD}' \approx v_{IFB(AVG)}$

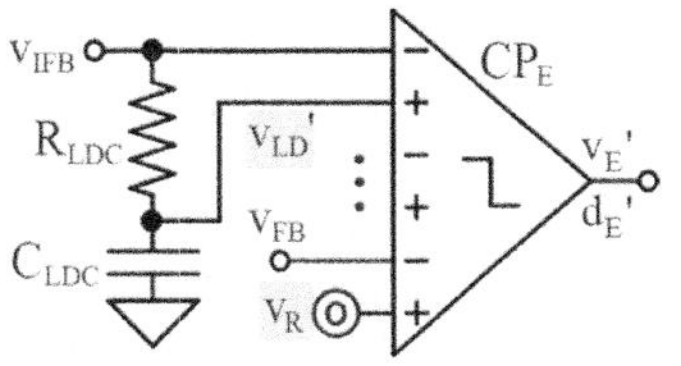

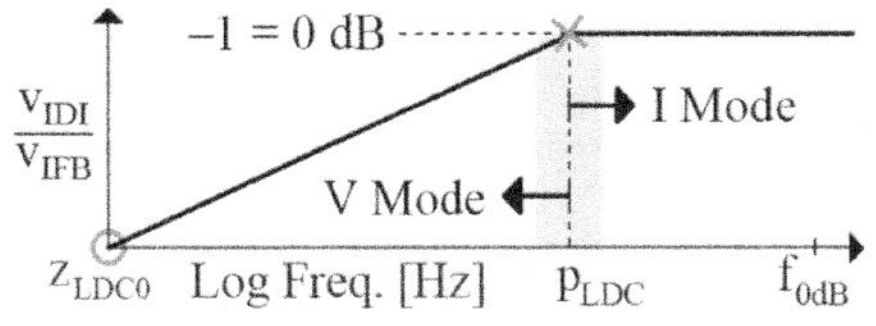

In Practice:

p_{LDC} can near p_{LC}

(if z_{DO} = Absent)

Operation: $f_O << p_{LDC}$ → $\Sigma v_{ID} = v_R \ldots + v_{IDI} \approx v_R \ldots + 0$ ∴ V Mode

$f_O >> p_{LDC}$ → $\Sigma v_{ID} = v_R \ldots + v_{IDI} \approx v_R \ldots - v_{IFB}$ ∴ I Mode

Stability: Reach f_{0dB} as current mode ∴ $p_{LDC} \approx \dfrac{1}{2\pi R_{LDC} C_{LDC}} << f_{0dB}$

ii. Load-Compensated Contractions

Load Compensation:

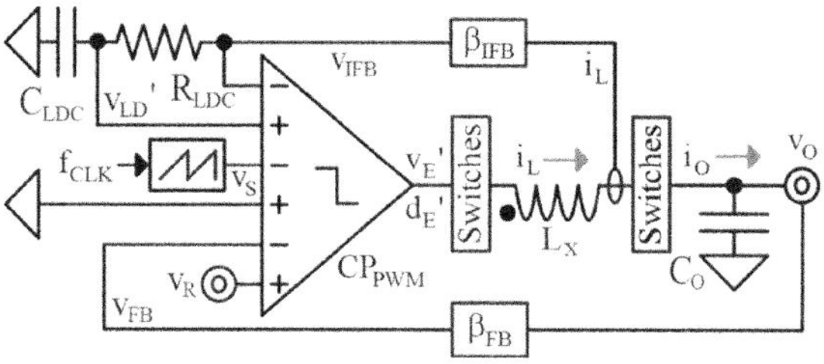

Add v_{LD}' to v_R

PWM Loop:

$\Sigma v_{ID} = v_R - v_{FB} - v_{IFB} - v_S + v_{LD}'$

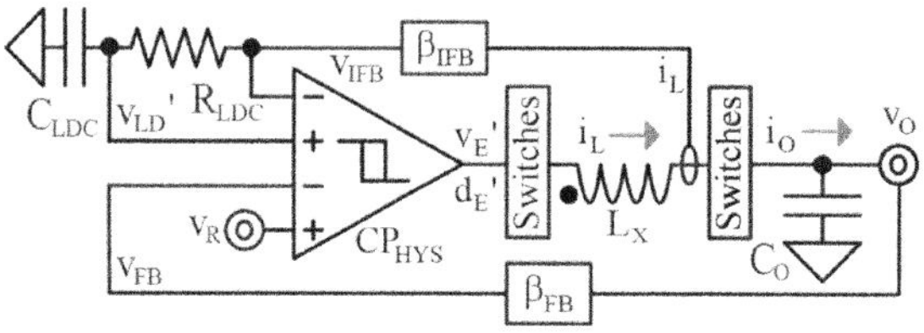

Hysteretic Loop:

$\Sigma v_{ID} = v_R - v_{FB} - v_{IFB} + v_{LD}'$

Constant-t_{OFF} Peak Loop:

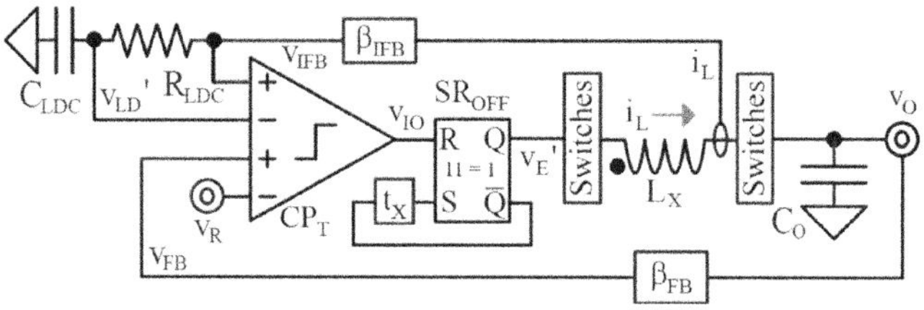

SR_{OFF} inverts $\therefore$

$\Sigma v_{ID} = - v_R + v_{FB} + v_{IFB} - v_{LD}'$

Example: Determine v_{IOS}, v_{VOS}, $v_{FB(AVG)}$, & β_{FB} from previous PWM example.

Solution: $v_{VOS} = v_{IOS} = 335$ mV from previous example

$$v_{FB(AVG)} = v_R - v_{VOS} = 1.2 - 335m = 865\text{ mV}$$

$$\beta_{FB} \equiv \frac{\overline{v_{FB(AVG)}}}{v_{O(AVG)}} = \frac{865m}{1.80} = 48\%$$

$$v_{O(AVG)} = \frac{v_{FB(AVG)}}{\beta_{FB}} = \frac{865m}{48\%} = 1.80\text{ V} \rightarrow \text{No loading effect}$$

7.5. Oscillating Voltage-Mode Bucks: A. Resistive Capacitor

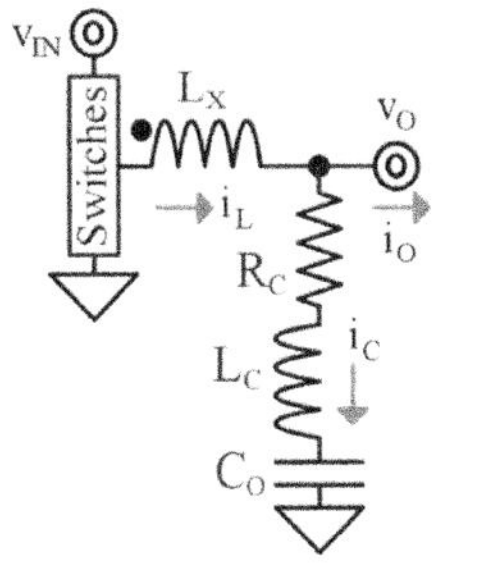

Currents: $i_O = i_{L(DC)} = i_{L(AVG)}$ $\qquad i_C = i_{L(AC)} = \Delta i_L$

Output: $v_{O(DC)} = v_{O(AVG)} = v_{C(AVG)} = v_{C(DC)}$

$v_{O(AC)} = v_R + v_L + v_{C(AC)}$

$v_R = \Delta i_L R_C$ → Triangular

$v_L = L_C\left(\frac{di_L}{dt_{E/D}}\right)$ → Pulsing

$v_C = \int \frac{i_{L(AC)}}{C_O} dt$ → Parabolic

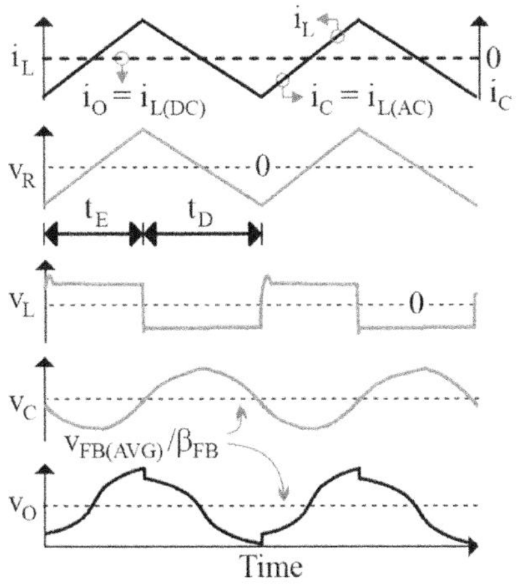

Resistive C_O: $\Delta v_O = v_{O(AC)} \approx v_R = \Delta i_L R_C$

→ Rises & falls with i_L

Comparator Loops:

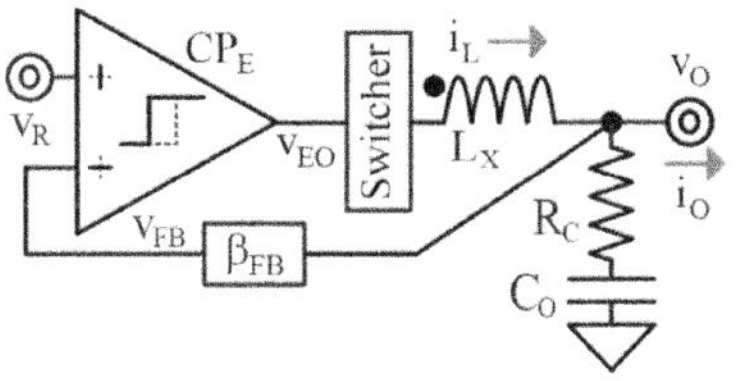

v_O rises across t_E & falls across t_D

→ Like v_{IFB} in hyst./timed loops

∴ Close similar loops (oscillators) with v_O → $\Delta v_{FB} = \Delta v_O \beta_{FB} \approx \Delta i_L R_C \beta_{FB}$

Hysteretic: $\Delta v_{FB} = \Delta v_T + v_{POS}^{+} + v_{POS}^{-} \approx \Delta v_T + t_P\left(\frac{v_E + v_D}{L_X}\right)R_C\beta_{FB}$

$$t_{SW} = \frac{t_E}{d_E} = \left(\frac{\Delta i_L}{d_E}\right)\left(\frac{dt_E}{di_L}\right) = \left(\frac{\Delta v_{FB}}{\beta_{FB}R_C}\right)\left(\frac{L_X}{v_E \parallel v_D}\right) = f(\Delta v_T, R_C, t_P, v_{E/D}, L_X)$$

v_{HOS} on $v_{FB(AVG)} = \frac{v_{POS}^{+} - v_{POS}^{-}}{2} \approx t_P\left(\frac{v_E - v_D}{2L_X}\right)R_C\beta_{FB}$ $\qquad d_E = \frac{v_D}{v_E + v_D}$

(– in v_{VOS})

Timed: $\Delta v_{FB} = t_{E/D}\left(\frac{dv_{FB}}{dt_{E/D}}\right)$ $\qquad v_{POS}^{\pm} = t_P'\left(\frac{dv_{FB}}{dt_{E/D}}\right)$ $\qquad \frac{dv_{FB}}{dt_{E/D}} = \left(\frac{v_{E/D}}{L_X}\right)R_C\beta_{FB}$

B. RC Filter

Comparator Loops:

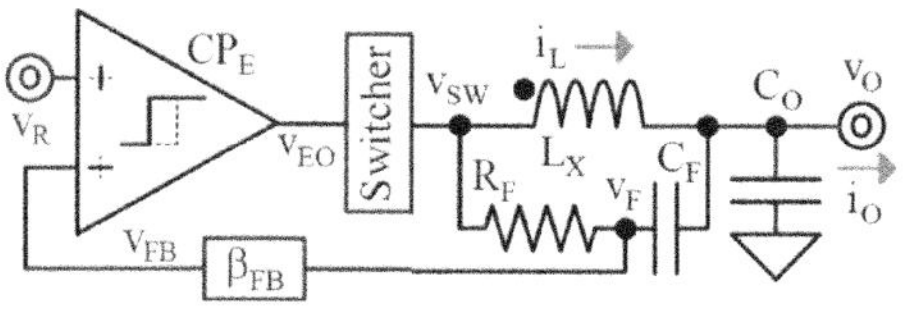

Low f_O → L_X = Short, C_F = Open ∴ $v_{F(AVG)} = v_{SW(AVG)} = v_{O(AVG)}$

C_F shorts below f_{SW} → $f_F = \frac{1}{2\pi R_F C_F} << f_{SW}$

∴ $v_R \approx v_L$ at f_{SW} (across t_{SW}) → $\frac{v_R}{R_F} \approx \frac{v_{SW(E/D)} - v_{F(AVG)}}{R_F} \approx \frac{v_{E/D}}{R_F}$ slews C_F up/down

$$\Delta v_F = \left(\frac{dv_F}{dt_{E/D}}\right) t_{E/D} = \left(\frac{i_C^{\pm}}{C_F}\right) t_{E/D} \approx \left(\frac{v_{E/D}}{R_F C_F}\right) t_{E/D} \rightarrow v_F \text{ rises/falls across } t_{E/D}$$

Note: $L_X C_O$ suppresses Δv_O, $R_F C_F$ suppresses Δv_{FB} → Δv_O can be lower than Δv_F

With low R_C: $\Delta v_O \approx \left(\frac{i_{C(\pm AVG)}}{C_O}\right)\left(\frac{t_{SW}}{2}\right) = \left(\frac{0.5\Delta i_L}{2C_O}\right)\left(\frac{t_{SW}}{2}\right) = \frac{\Delta i_L}{8 f_{SW} C_O}$

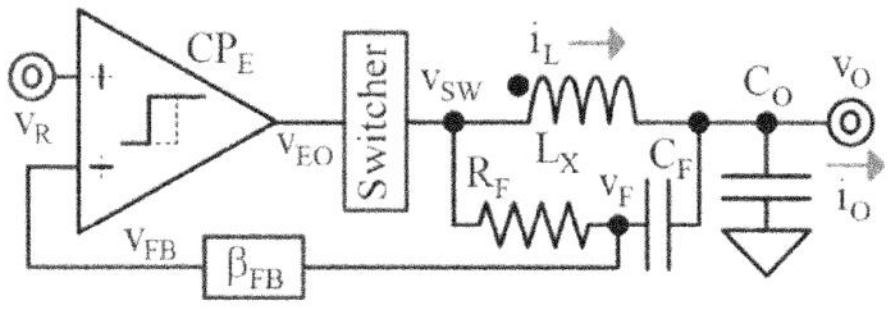

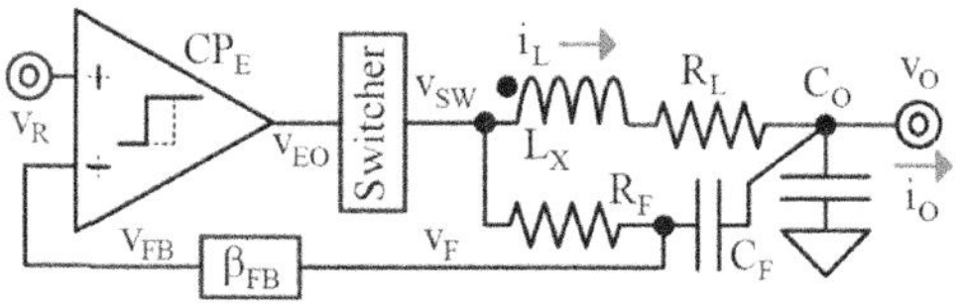

Hysteretic: $\Delta v_{FB} = \Delta v_F \beta_{FB} \approx \Delta v_T + t_P\left(\frac{dv_F}{dt_E} + \frac{dv_F}{dt_D}\right)\beta_{FB} \approx \Delta v_T + t_P\left(\frac{v_E + v_D}{R_F C_F}\right)\beta_{FB}$

$$t_{SW} = \frac{t_E}{d_E} = \Delta v_F\left(\frac{dt_E}{dv_F}\right)\left(\frac{v_E + v_D}{v_D}\right) \approx \left(\frac{\Delta v_{FB}}{\beta_{FB}}\right)\left(\frac{R_F C_F}{v_E \| v_D}\right)$$

v_{HOS} on $v_{FB(AVG)} = \frac{v_{POS}^+ - v_{POS}^-}{2} \approx \left(\frac{t_P}{2}\right)\left(\frac{dv_F}{dt_E} - \frac{dv_F}{dt_D}\right)\beta_{FB} \approx t_P\left(\frac{v_E - v_D}{2R_F C_F}\right)\beta_{FB}$

(– in v_{VOS})

Timed: $\Delta v_{FB} = \Delta v_F \beta_{FB} \approx t_{E/D}\left(\frac{v_{E/D}}{R_F C_F}\right)\beta_{FB}$ $\quad v_{POS}^{\pm} = t_P'\left(\frac{dv_F}{dt_{E/D}}\right)\beta_{FB} = t_P'\left(\frac{v_{E/D}}{R_F C_F}\right)\beta_{FB}$

Loading: R_L reduces $v_{O(AVG)}$ → Adds $v_{LD} = i_O R_{L(DC)} \beta_{FB}$ to v_{VOS}

Voltage-Mode Voltage Loops ≡ Voltage-Squared Loops

Use filtered buck as current loop, but as a voltage buffer → $A_{V1} = \frac{v_O}{v_{EO}}$

→ β_{FB} scales v_O to v_{FB}

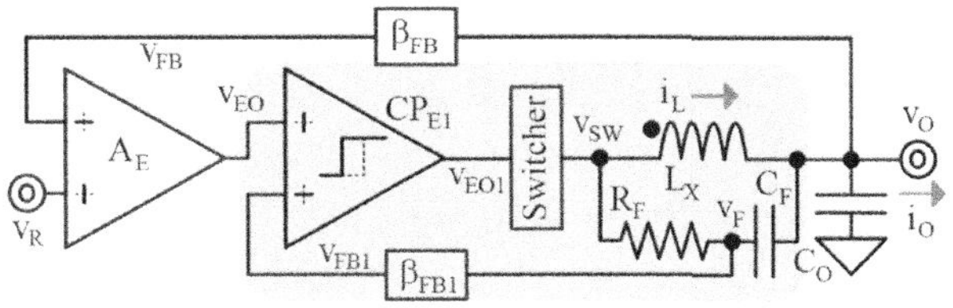

A_E compares v_{FB} to v_R

Inner loop buffers v_{EO} so $v_O = v_{EO}A_{V1} \approx \frac{v_{EO}}{\beta_{FB1}}$

Loop Offset: $v_{LOS} = v_{EO} \approx v_{FB1(AVG)} = v_{F(AVG)}\beta_{FB1} = v_{SW(AVG)}\beta_{FB1} \approx v_{O(AVG)}\beta_{FB1}$

$$v_{VOS} \equiv v_R - v_{FB(AVG)} = \frac{v_{EO}}{A_E} = \frac{v_{IOS} + v_{LOS} + v_{LD}}{A_E}$$

In Practice: $v_{O(AVG)} > 1$ $\quad R_L = \text{Low}$ $\quad \therefore$ $\quad v_{LOS} >> v_{IOS}, v_{LD}$

β_{FB1} centers $v_{IOS} + v_{LD}$, A_{V1} responds quickly, A_E suppresses $v_{IOS} + v_{LOS} + v_{LD}$

7.6. Summary

PWM Loops → p_{E1} averages response $\therefore$ t_R = Multiple t_{SW} cycles

→ Sawtooth offsets $v_{R/IR}$

Hysteretic Loops → Shortest possible t_R with f_{SW} sensitivity

Constant $t_{E/D}$ Loops → Moderate t_R with moderate f_{SW} sensitivity

Constant t_{SW} Loops → Short t_R with no f_{SW} sensitivity + slope compensation

Hys./Timed Loops → CP delay projects v_{IR} offset

Peak/Valley Loops → Half v_{IFB} ripple offsets v_{IR}

i_L-Mode v_O Loops → Load offsets v_R

Contractions → Lose offset suppression $\therefore$ Add v_{LD}' to v_R

Chapter 8. Building Blocks

8.1. Current Sensors

8.2. Voltage Sensors

8.3. Digital Blocks

8.4. Comparator Blocks

8.5. Timing Blocks

8.6. Switch Blocks

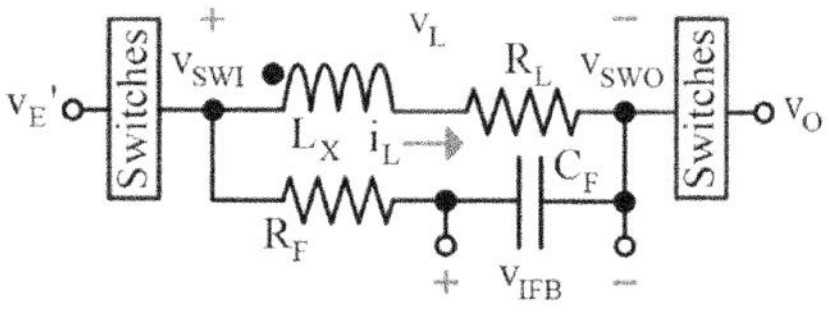

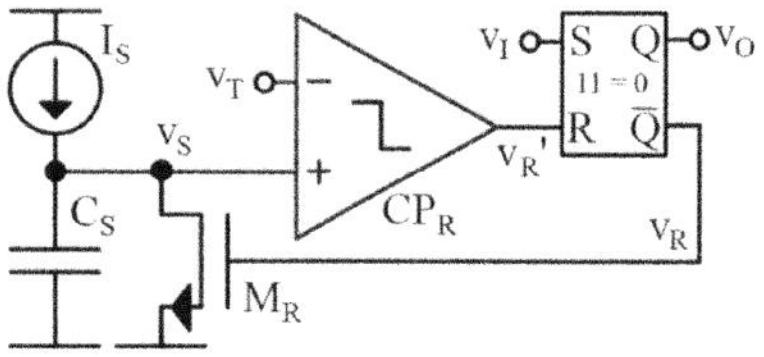

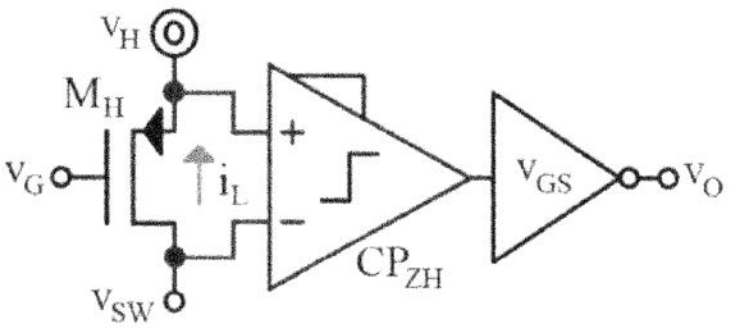

8.1. Current Sensors: A. Series Resistance

Insert R_S:

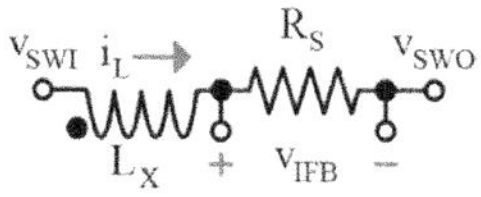

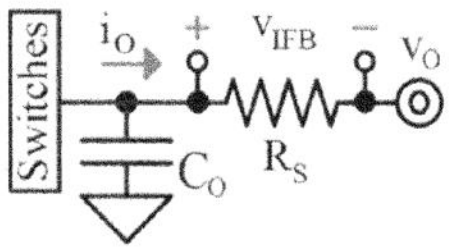

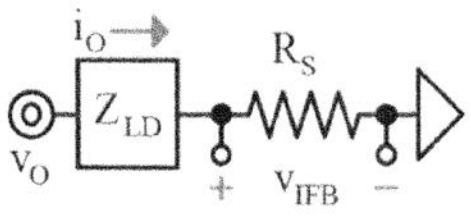

Drawback: R_S burns P_R ∴ Use low R_S → v_{IFB} = Low

$$\beta_{IFB} = \frac{v_{IFB}}{i_{L/O}} = R_S$$

∴ A_{IE} or CP_{IE} needs more P_Q to discern v_{IFB} from noise

Sense R_{DS}: Already in the conduction path → No added P_R

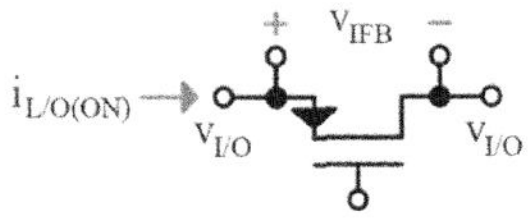

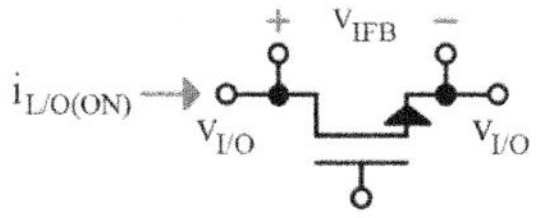

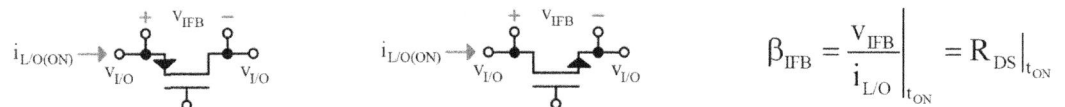

$$\beta_{IFB} = \left.\frac{v_{IFB}}{i_{L/O}}\right|_{t_{ON}} = R_{DS}\big|_{t_{ON}}$$

Drawbacks: Senses part of $i_{L/O}$ → Reconstruct v_{IFB} (sense with $R_{DS(E)}$ & $R_{DS(D)}$)

R_{DS} = Low ∴ A_{IE} or CP_{IE} needs more P_Q to discern v_{IFB} from noise

Sense R_L: Already in the conduction path → No added P_R

Low-Pass RC Filter: Average v_{SW}'s

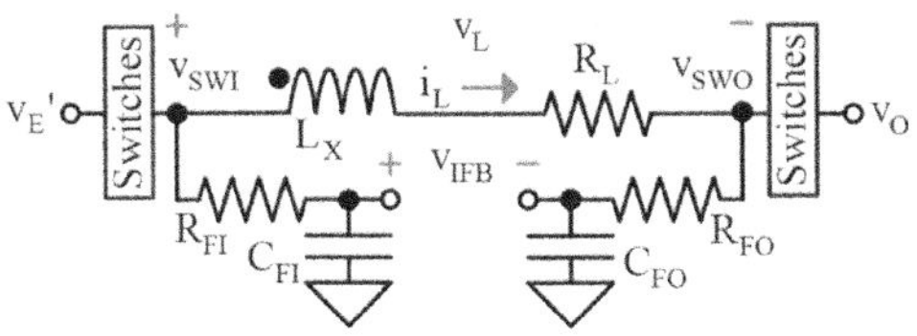

$$\beta_{IFB} = \frac{v_{IFB}}{i_{L(AVG)}} = \frac{v_{SWI(AVG)} - v_{SWO(AVG)}}{i_{L(AVG)}}$$

$$\approx \frac{v_{L(DC)}}{i_{L(AVG)}} = \frac{i_{L(AVG)} R_{L(DC)}}{i_{L(AVG)}} = R_{L(DC)}$$

Drawbacks: R_L = Low ∴ A_{IE} or CP_{IE} needs more P_Q to discern v_{IFB} from noise

v_{SW}'s swing $v_{IN/O}$ → C_F's should suppress noise ∴ $Z_C \ll R_F$

$$\rightarrow \quad v_n \approx \frac{i_{FI/O}{}^{\pm} t_{E/D}}{C_{FI/O}} \approx \left(\frac{\Delta v_{SWI/O}}{R_{FI/O}}\right)\left(\frac{d_{E/D} t_{SW}}{C_{FI/O}}\right) \approx \left(\frac{v_{IN/O} t_{SW}}{R_{FI/O} C_{FI/O}}\right)\left(\frac{v_{D/E}}{v_E + v_D}\right) \ll v_{L(DC)}$$

↑ Design

Notes: β_{IFB} senses $i_{L(AVG)}$ → Low-bandwidth translation

Buck → No $R_{FO}C_{FO}$ needed Boost → No $R_{FI}C_{FI}$ needed

Bypass RC Filter: L_X filters v_L into i_L like C_F filters v_L into v_{IFB}

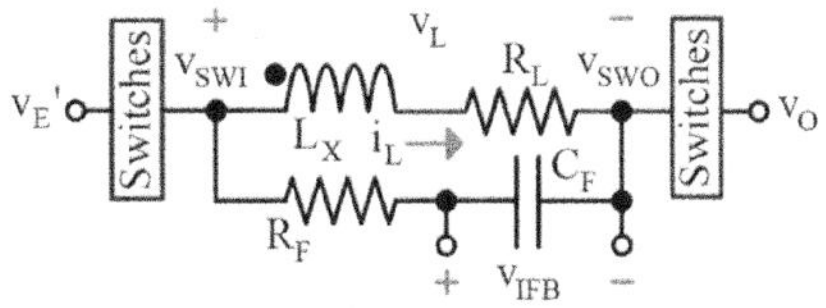

$$i_L = \frac{v_L}{sL_X + R_L} = \frac{v_L/R_L}{sL_X/R_L + 1}$$

$$v_{IFB} = \left(\frac{v_L}{1/sC_F + R_F}\right)\left(\frac{1}{sC_F}\right) = \frac{i_L\left(sL_X + R_L\right)}{1 + sR_F C_F} = i_L R_L \left(\frac{1 + sL_X/R_L}{1 + sR_F C_F}\right) = i_L R_L = i_L \beta_{IFB}$$

↑

$$\text{If } f_{RC} = \frac{1}{2\pi R_F C_F} \equiv f_{RL} = \frac{R_L}{2\pi L_X}$$

Feature: β_{IFB} senses i_L across f_O → High-bandwidth translation

Drawbacks: R_L = Low ∴ A_{IE} or CP_{IE} needs more P_Q to discern v_{IFB} from noise

Skin effect increases R_L → $\beta_{IFB} \propto f_O$

v_{SW}'s swing $v_{IN/O}$ → Wide ICMR ($v_{P/N}$ range) for A_{IE} or CP_{IE}

R_FC_F shunts/bypasses L_XR_L past f_B: When sL_X overcomes R_F

$$R_F + \frac{1}{sC_F} = R_F + \frac{R_L R_F}{sL_X} = \left(\frac{R_F}{sL_X}\right)(sL_X + R_L) \le sL_X + R_L \text{ past } f_B = \frac{R_F}{2\pi L_X}$$

$$f_{RC} \equiv f_{RL}$$

L_X in μH's & R_F in kΩ's → $f_B \approx$ GHz ∴ Little effect near f_{0dB}

f_O >> Unmatched f_{RL} & f_{RC}: R_L "shorts" C_F shorts

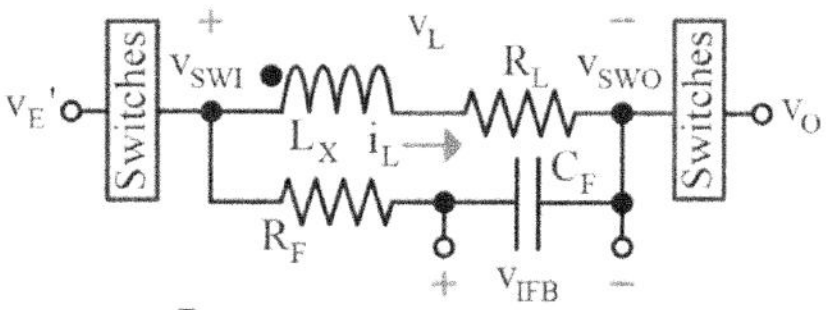

$$\beta_{IFB(MF)} = \frac{v_{IFB}}{i_L}\Bigg|_{f_O >> f_{RC}}^{f_O >> f_{RL}} \approx \frac{i_C Z_C}{i_L} \approx \left(\frac{v_L}{R_F}\right)\left(\frac{Z_C}{i_L}\right) \approx \frac{i_L sL_X}{R_F sC_F i_L} = \frac{L_X}{R_F C_F} > R_L \text{ if } f_{RL} < f_{RC}$$

$$\downarrow$$

$$z_L > p_C$$

Low f_O: L_X shorts R_F "shorts" ∴ $\beta_{IFB(LF)} = \frac{v_{IFB}}{i_L}\Bigg|_{f_O << f_{RC}}^{f_O << f_{RL}} \approx \frac{i_L R_L}{i_L} = R_L$

$$f_B': \quad R_F + \frac{1}{sC_F} \equiv R_F + \frac{\beta_{IFB(MF)} R_F}{sL_X} \le sL_X + R_L \text{ past } f_B' = f_B\left(\frac{sL_X + \beta_{IFB(MF)}}{sL_X + R_L}\right) > f_B$$

Example: Determine C_F, $v_{IFB(AVG)}$, & f_B so $\beta_{IFB} = R_L$ for a buck–boost when R_F = 500 kΩ, L_X = 10 μH, R_L = 250 mΩ, $i_{L(AVG)}$ = 300 mA.

Solution:

$$f_{RC} = \frac{1}{2\pi R_F C_F} = \frac{1}{(500k)C_F} \equiv f_{RL} = \frac{R_L}{2\pi L_X} = \frac{250m}{10\mu} \quad \therefore \quad C_F = 80 \text{ pF}$$

$$v_{IFB(AVG)} = i_{L(AVG)} R_L = (300m)(250m) = 75 \text{ mV}$$

$$f_B = \frac{R_F}{2\pi L_X} = \frac{500k}{2\pi(10\mu)} = 8.0 \text{ GHz} \rightarrow \text{Very high}$$

B. Sense Transistor

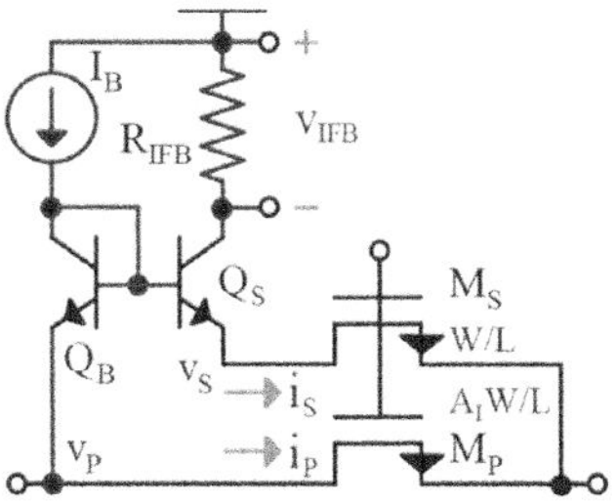

Sense FET: M_S mirrors M_P → Same v_{GS}'s

Q_B & Q_S match their v_D's

$$\therefore \quad \beta_{IFB} = \left.\frac{v_{IFB}}{i_{L/O}}\right|_{t_{ON}} \approx \left.\frac{(i_{L/O}/A_I)R_{IFB}}{i_{L/O}}\right|_{t_{ON}} = \left.\frac{R_{IFB}}{A_I}\right|_{t_{ON}}$$

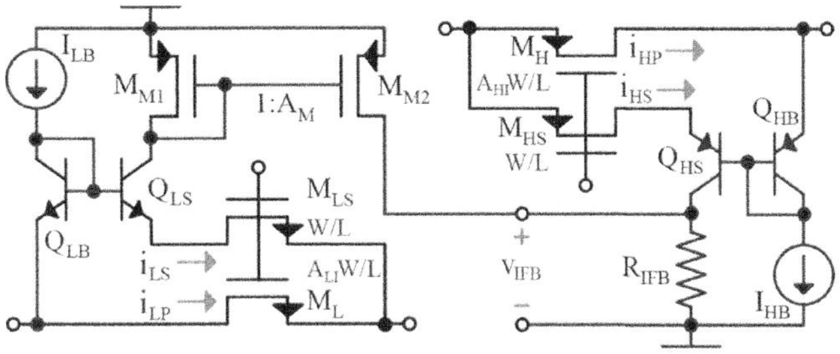

v_D Error:

$$v_E = v_S - v_D = \Delta v_{BE} = V_t \ln\frac{i_S}{I_B}$$

Reconstruct i_L:

$$\beta_{IFB} = \frac{v_{IFB}}{i_L} \approx \frac{(i_{LS}A_M + i_{HS})R_{IFB}}{i_L} \approx \left(\frac{i_{L(L)}A_M}{A_{LI}} + \frac{i_{H(L)}}{A_{HI}}\right)\left(\frac{R_{IFB}}{i_L}\right) = \frac{R_{IFB}}{A_I}$$

Fix Error: $i_B = i_S$

Looped Sense FET:

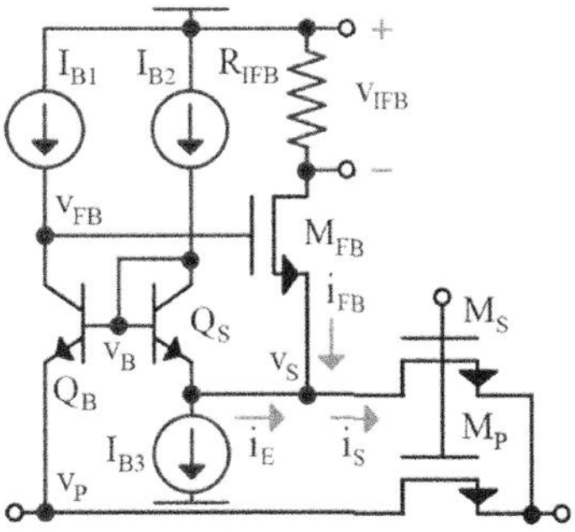

M_S mirrors M_P → Same v_{GS}'s

$I_{B1} = I_{B2}$ → Q_B & Q_S match their v_D's

v_S compares i_S & i_{FB}

Q_S level-shifts v_S, Q_B amplifies v_B,

& M_{FB} buffers v_{FB} ∴ A_{LG} is high → $i_{FB} \approx i_S$

$R_S << R_{FB}$ → $p_{FB} << p_S$ ∴ Stable

Tradeoff: f_{0dB} limits t_R → Slower

Reconstruct i_L: Mirror i_{FB} into the R_{IFB} of a high-side sensor.

C. Comparison

	Series Resistances					Sense FETs	
	R_S	R_{DS}	Filtered R_L			Basic	Looped
			Low Pass	Tuned	Untuned		
Input	$i_{O/L}$	$i_{O/L(E/D)}$	$i_{L(DC)}$	i_L	$i_{L(DC/AC)}$	$i_{O/L(E/D)}$	
Gain	R_S	R_{DS}	R_L	R_L	$\frac{L_X}{R_F C_F}$	$\frac{R_{IFB}}{A_I}$	
Power	$P_R + P_Q$	P_Q	P_Q	P_Q	P_Q	P_Q	
Sensitivity	T_J & Fabrication Runs $R_L \propto f_O$					Mismatch	
						Nonlinear	Linear

Accuracy: Transistor mismatch is usually less sensitive than resistor tolerance.

Tradeoff: Nonlinearity is usually less constraining than response time.

In Practice: Basic sense FETs often offer more favorable tradeoffs.

8.2. Voltage Sensors: A. Voltage Divider

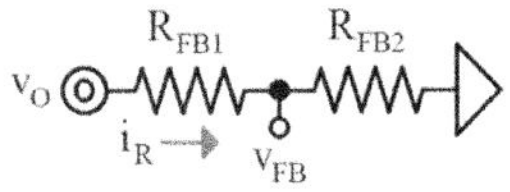

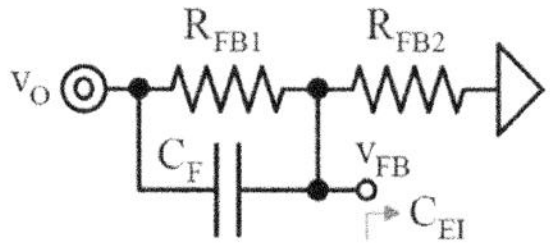

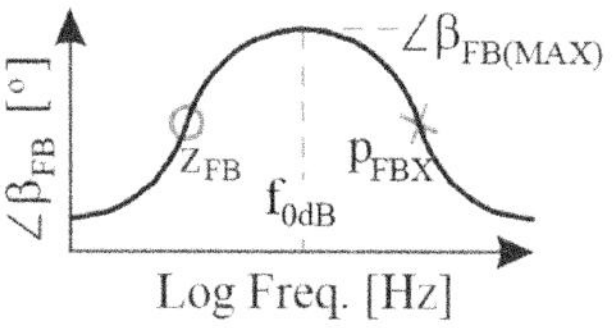

Basic: Usually, $v_O > v_R$

R's match & track well across process, T_J

$$\beta_{FB} \equiv \frac{v_{FB}}{v_O} = \frac{R_{FB2}}{R_{FB1} + R_{FB2}}$$

∴ Voltage dividers are usually accurate

B. Phase-Saving: z_{FB} before f_{0dB} recovers phase & p_{FBX} after f_{0dB} loses recovery.

C_F bypasses R_{FB1} $\qquad$ C_O shunts v_O past p_O → $R_{FB1} \| R_{FB2}$ i-limit C_F & C_{EI}

$$z_{FB} = \frac{1}{2\pi R_{FB1} C_F} \qquad p_{FBX} \approx \frac{1}{2\pi (R_{FB1} \| R_{FB2})(C_F + C_{EI})} \approx \frac{1}{2\pi (R_{FB1} \| R_{FB2}) C_F}$$

$$\Delta PM = \tan^{-1}\frac{f_{0dB}}{z_{FB}} - \tan^{-1}\frac{f_{0dB}}{p_{FBX}}$$ maxes when centered about f_{0dB} $\quad \therefore \quad \frac{f_{0dB}}{z_{FB}} \equiv \frac{p_{FBX}}{f_{0dB}}$

C. Voltage-Dividing Error Amplifier

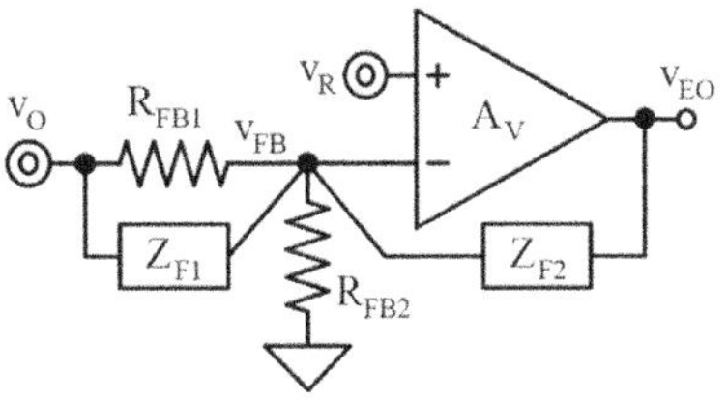

Goal: Combine voltage divider & error amp

Notes: $Z_{F1/2}$ alter $\frac{v_{EO}}{v_R}$ & $\frac{v_{EO}}{v_O}$ differently

$\frac{v_{EO}}{v_O}$ determines stability

FB only at higher f_O → Mixed Translation → $Z_F \equiv C_F$ (and series R_F)

Translation: $$A_F = \left[\frac{R_{FB2} \,||\, Z_{F2}}{(R_{FB1} \,||\, Z_{F1}) + (R_{FB2} \,||\, Z_{F2})}\right]\left(\frac{-A_{V0}}{1+s/2\pi p_A}\right) = -\beta_{FB0}A_{V0}A_X$$

$$A_\beta = \frac{-Z_{F2}}{R_{FB1} \,||\, Z_{F1}} \qquad \frac{v_{EO}}{v_O} \equiv \beta_{FB}A_E = A_F \,||\, A_\beta = -\beta_{FB0}A_{V0}A_S$$

↓

E.g.: A_S with C_{F1}:C_{F2}:R_{F2} → C_{F2} reduces A_β below A_F: p_{E1}, $A_{S0} = 0$ dB

flattens when R_{F2} current-limits C_{F2}: z_{E1}, rises when C_{F1} shunts R_{FB1}: z_{E2}.

8.3. Digital Blocks: A. Push–Pull Logic

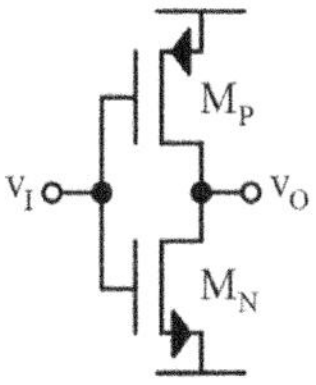

Inverter: Pull high with PFET Pull low with NFET

Threshold: Balances at v_T when MOS strengths match

Max noise immunity → $v_T \equiv \frac{V_{DD} - V_{SS}}{2}$

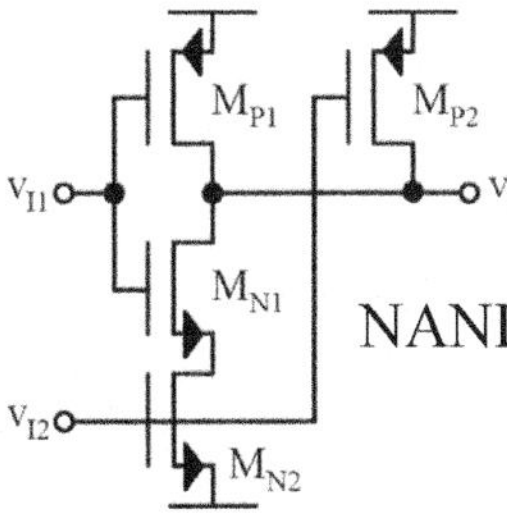

NAND

Design: Short t_P → Low C's

∴ Use L_{MIN}'s W_{MIN} for stronger device

Raise weaker's W so $i_P = i_N$ at $|v_{GS}| \equiv |v_{DS}| \equiv v_T$:

$$\left.\frac{i_P}{i_N}\right|_{v_T}^{v_{GS}>V_{T0}} \approx \frac{W_P L_N K_P'\left(v_T - |V_{TP0}|\right)^2(1+v_T\lambda_P)}{W_N L_P K_N'\left(v_T - V_{TN0}\right)^2(1+v_T\lambda_N)} \equiv 1$$

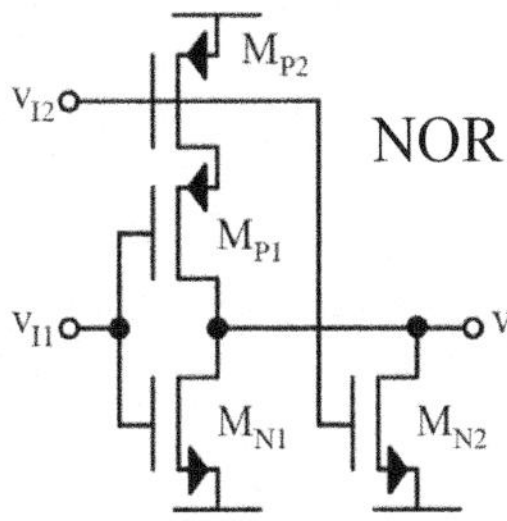

NOR

NAND: M_{N2} reduces v_{GSN1} ∴ Double $W_{N1} \equiv W_{N2}$

NOR: M_{P2} reduces v_{SGP1} ∴ Double $W_{P1} \equiv W_{P2}$

B. SR Flip Flops

S	R	Q
0	0	Q^{-1}
0	1	0
1	0	1
1	1	1

S	R	Q
0	0	Q^{-1}
0	1	0
1	0	1
1	1	0

Temporal:

S sets high → Switch to v_{DD}

R resets low → Switch to v_{SS}

0's hold Q^{-1} → C_H holds Q^{-1}

1's set/reset → $R_{S/R} << R_{R/S}$

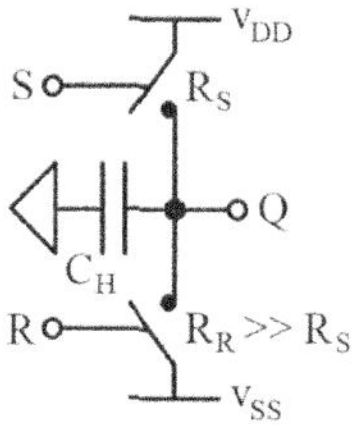

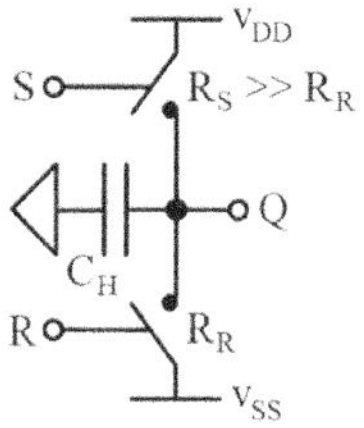

Digital:

S sets high → OR S

R resets low → OR R

0's hold Q^{-1} → + FB latches Q^{-1}

1's set/reset → Tap S/R OR

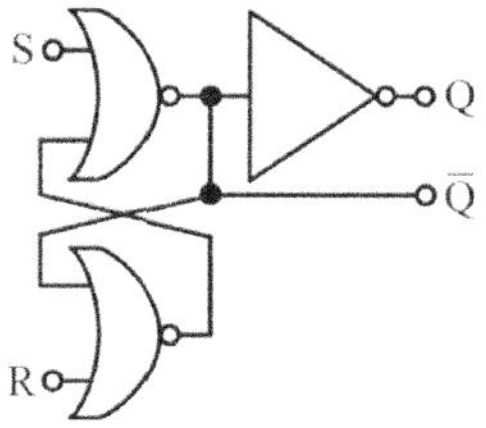

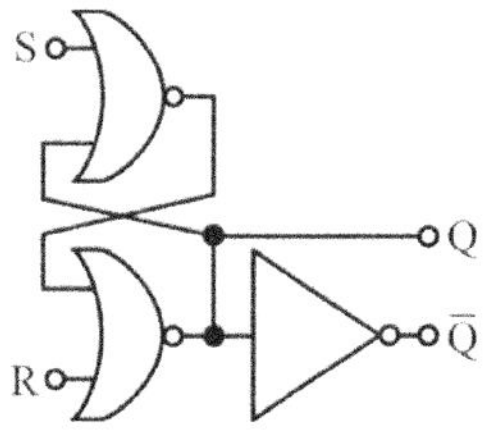

C. Gate Driver

Design: Large switches → High C_{GO} → Chain of increasingly larger inverters

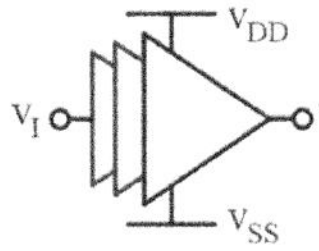

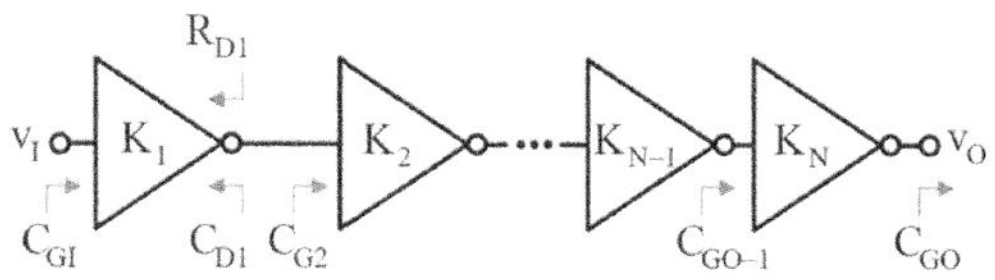

High total fan-out $F_O = \frac{C_{GO}}{C_{GI}} = \frac{C_{GI} f_o^{\ N}}{C_{GI}} = f_o^{\ N} = \left(\frac{C_G}{C_{G-1}}\right)^N = \frac{W_{GO}}{W_{GI}}$

$t_{P(MIN)}$: $t_{P1} \equiv t_{50\%}$ of $\Delta v_{O(MAX)} \approx 69\%\tau_{RC} = 69\% R_{D1} C_{O1}$ $f_o \equiv$ Inter-stage fan-out

$C_{O1} = C_{D1} + C_{G2} = C_{GI}(k_{SL} + f_o) \approx C_{GI}(1 + f_o)$

↓

Self-Loading Coefficient ≈ 1

$t_P = N t_{P1} \approx 69\% N R_{D1} C_{GI}(1 + f_o)$

$N(1 + f_o)$ bottoms when $f_o = 3.6$ → $t_{P(MIN)}$

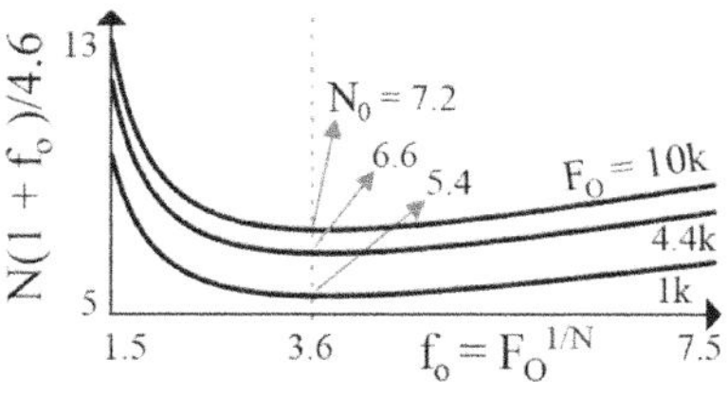

Gate-Charge Power:

$$P_G = \sum_{1}^{N} P_{G(K)} = P_{G1}\sum_{0}^{N-1} f_o^{\ k}$$

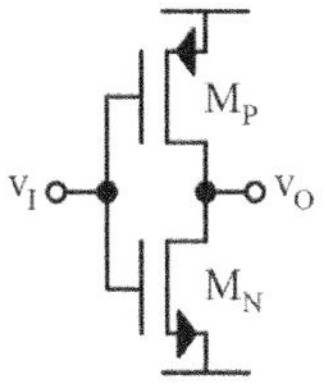

$$q_{G1} = C_{O1}\Delta v_{O(MAX)} = C_{O1}(v_{DD} - v_{SS})$$

$$P_{G1} = (v_{DD} - v_{SS})q_{G1}f_{SW} = C_{O1}(v_{DD} - v_{SS})^2 f_{SW} \propto C_{O1}$$

Shoot-Through Power: Balance $M_{P/N}$ at $v_T \equiv 0.5(v_{DD} - v_{SS})$

v_{GS}'s, v_{DS}'s max at v_T $\therefore$ $i_{P/N}$ max at v_T

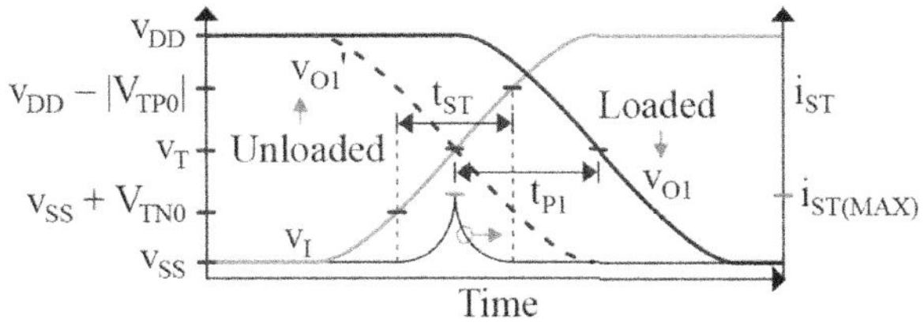

$$i_T \approx \left(\frac{W_N}{L_N}\right)\left(\frac{K_N'}{2}\right)\left(v_T - V_{TN0}\right)^2\left(1 + v_T\lambda_N\right)$$

M_P & M_N conduct between V_{T0}'s:

$$P_{ST1} \propto i_{ST1(MAX)} \propto \frac{1}{t_{P1}} \propto \frac{1}{1+f_o} \approx \left(\frac{30\%}{1+f_o}\right)i_{T1}$$

$$t_{ST} = t_{RC}\Big|_{v_{SS}+V_{TN0}}^{v_{DD}-|V_{TP0}|} \approx \tau_{RC}\ln\left[\frac{\left(v_{DD} - v_{SS}\right) - \left(v_{DD} - |V_{TP0}|\right)}{\left(v_{DD} - v_{SS}\right) - \left(v_{SS} + V_{TN0}\right)}\right]^{-1}$$

$$P_{ST1} \approx \left(\frac{i_{ST1(MAX)}}{3}\right)\left(v_{DD} - v_{SS}\right)\left(\frac{2t_{ST}}{t_{SW}}\right) \propto i_{T1} \propto W_{N/P1} \rightarrow C_{O1} \qquad P_{ST} = \sum_{1}^{N} P_{ST(K)} \approx P_{ST1}\sum_{0}^{N-1} f_o^{\ k}$$

Example: Determine N, t_P, P_G, & P_{ST} when $v_{DD} = 4$ V, $f_{SW} = 1$ MHz, $W_{N1} = 3$ µm,

$W_{GO} = 100$ mm, L's = 250 nm, $K_N' = 200$ µA/V^2, $\lambda_{N/P} = 5\%$,

$V_{TN0} = 500$ mV, $V_{TP0} = -700$ mV, $L_{OL} = 30$ nm, $C_{OX}'' = 6.9$ fF/µm^2.

Solution: $v_T = 2$ V when $W_{P1} = 6.7W_{N1} = 20$ µm

$$C_{GI} = C_{GN} + C_{GP} = C_{GN} + 6.7C_{GN} = 7.7C_{GN} = 7.7W_{N1}L_{N1}C_{OX}'' = 40 \text{ fF}$$

$$F_O = \frac{C_{GO}}{C_{GI}} = \frac{W_{GO}}{7.7W_{GN}} = 4.4\text{k} = f_o^{\ N_0} \equiv 3.6^{N_0} \quad \rightarrow \quad N_0 = 6.6 \quad \therefore \quad N \equiv 7,\ f_o = 3.3$$

$$R_{D1} \approx \frac{L_{N1} - 2L_{OL}}{W_{N1}K_N'\left(v_T - V_{TN0}\right)} = 210\ \Omega$$

$$C_{O1} \approx C_{GI}(1 + f_o) = 170 \text{ fF} \qquad t_P \approx 69\% N R_{D1} C_{O1} = 170 \text{ ps}$$

$$P_{G1} \approx C_{O1}v_{DD}^2 f_{SW} = 2.7\ \mu\text{W} \qquad P_G = P_{G1}\sum_{0}^{N-1} f_o^{\ k} = 5.0 \text{ mW}$$

$$i_{T1} \approx \left(\frac{W_N}{L_N - 2L_{OL}}\right)\left(\frac{K_N'}{2}\right)(v_T - V_{TN0})^2(1 + v_T\lambda_N) = 4\text{ mA}$$

$$\tau_{RC} \approx R_{D1}C_{O1} = 36\text{ ps} \qquad t_{ST} \approx \tau_{RC}\ln\left(\frac{v_{DD} - V_{TN0}}{|V_{TP0}|}\right) = 58\text{ ps}$$

$$P_{ST1} \approx \left[\frac{30\%i_{T1}}{3(1+f_o)}\right]v_{DD}\left(\frac{2t_{ST}}{t_{SW}}\right) = 43\text{ nW} \qquad P_{ST} \approx P_{ST1}\sum_0^{N-1} f_o^{\,k} = 80\ \mu\text{W}$$

Notes: $P_{ST} = 1.6\%P_G$ → $P_{DRV} \approx P_G$ when $f_o \geq 3.6$

$t_P \propto N(1 + f_o)$ $\quad P_{DRV} \propto \sum_0^{N-1} f_o^{\,k}$ → P_{DRV} = More sensitive to N

$t_P << t_{SW}$ $\quad\therefore\quad$ t_P = Less significant/sensitive → Limit N to 3 or 5

E.g.: $N \equiv 4$ → $t_P \approx 210$ ps (+24%) $\quad P_G \approx 3.5$ mW (–30%) $\quad P_{ST} \approx 26\ \mu$W

D. Dead-Time Delay

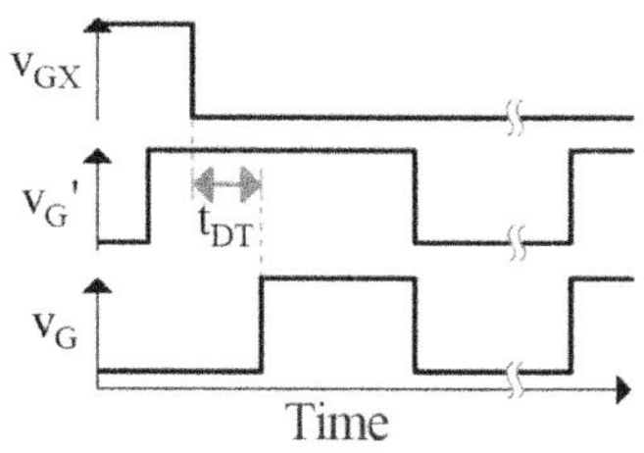

Problem: Adjacent switches can ground $v_{IN/O}$

Fix: Insert t_{DT} between conduction periods

But only when closing switches

∴ Disable when v_{GX} = High

→ Force v_G low

v_G rises t_{DT} after v_{GX} falls

→ Delay v_{GX} with RC or inverter chain

v_G falls with v_G' → Force v_G low

v_G rises with v_G' if v_{GX} is already low

→ $v_G' \neq$ Delayed

Active High

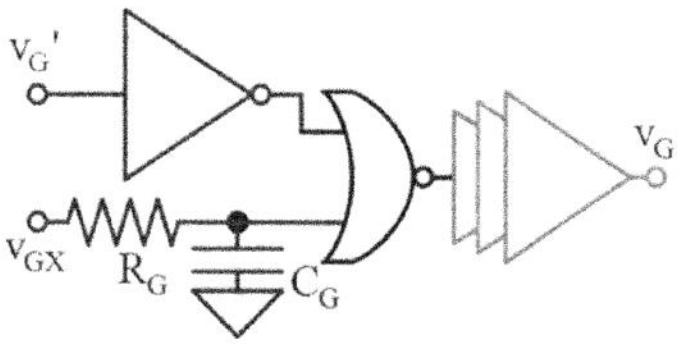

Active Low

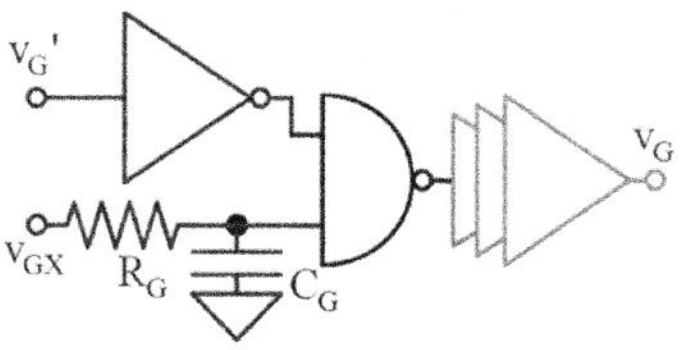

8.4. Comparator Blocks

Comparator = Un-stabilized amplifier → A_{V0}, p_{BW}, R_O, ...

Hysteretic: Set when v_{ID} climbs over $v_{T(HI)}$ Hold when $v_{T(LO)} < v_{ID} < v_{T(HI)}$

Reset when v_{ID} falls under $v_{T(LO)}$ Impossible: $v_{T(HI)} < v_{ID} < v_{T(LO)}$

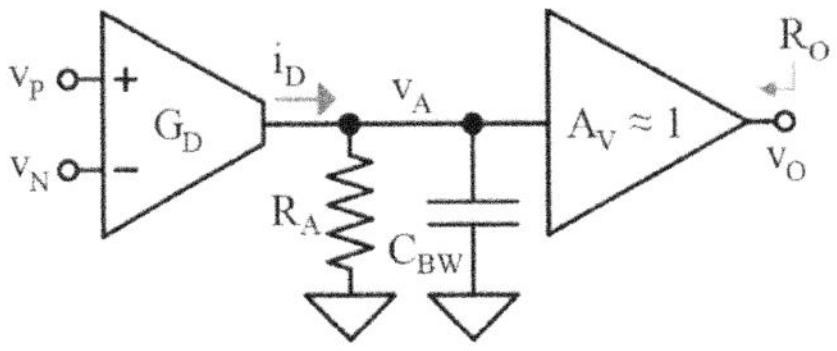

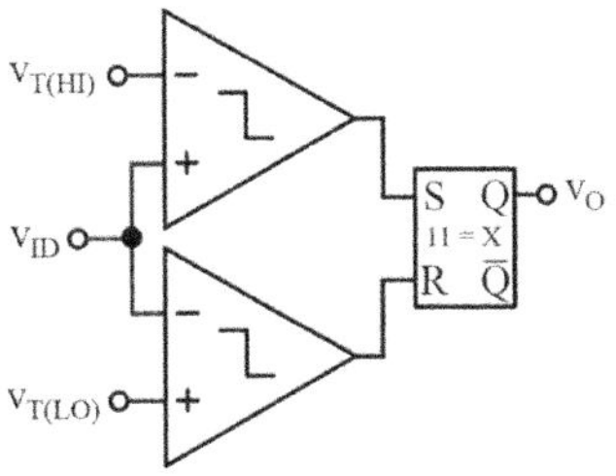

Summing Comparators: Parallel G_D's add currents

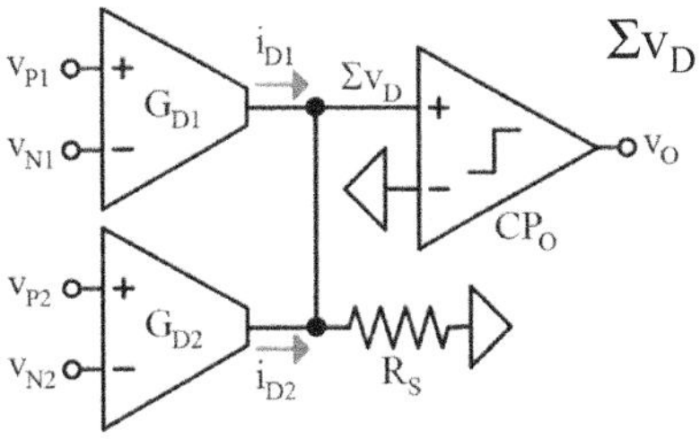

$\Sigma v_D = (v_{ID1} + v_{ID2})G_D R_S$

Hysteretic →

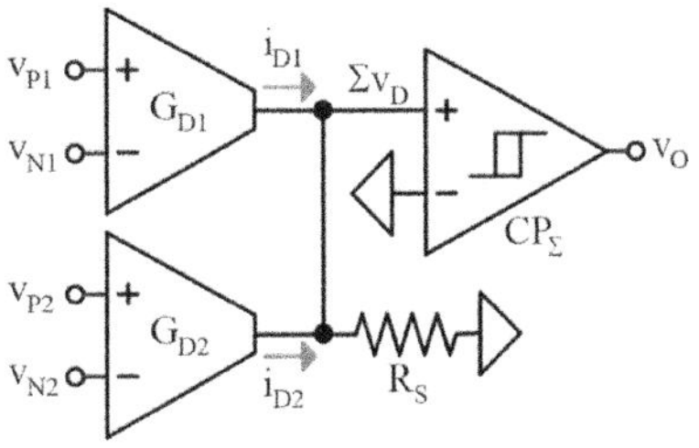

8.5. Timing Blocks

Clocked Sawtooth Generator:

I_S into C_S ramps v_S

f_{CLK} pulse resets ramp

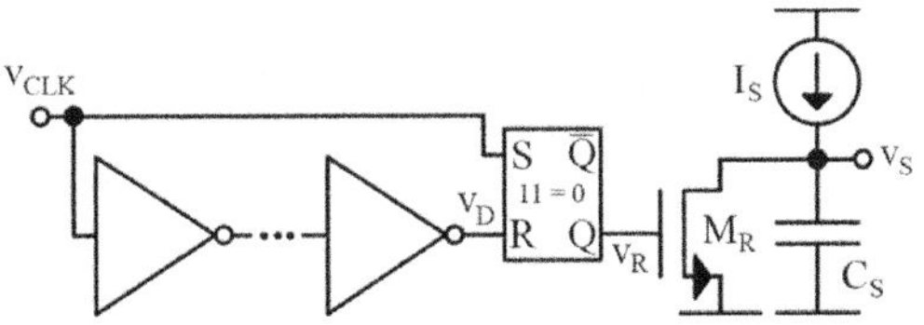

Design:

Δv_S's reset time $t_R \approx t_{90\%}$ of $\Delta v_S = 2.3 R_R C_S$ < Inverters' delay t_I

$$\Delta v_R \text{ into } C_{GD} \text{ shifts } v_{S(LO)} \approx v_{SS} - \Delta v_R\left(\frac{C_{GD}}{C_{GD}+C_S}\right) \approx v_{SS} - \left(v_{DD} - v_{SS}\right)\overbrace{\left(\frac{0.5C_{CH}}{C_S}\right)}^{\text{Triode } C_{GD}}$$

$$v_S \text{ rises across } t_{CLK} - t_I \quad \therefore \quad v_{S(HI)} \approx v_{S(LO)} + \left(\frac{I_S}{C_S}\right)\left(t_{CLK} - t_I\right)$$

Sawtooth Oscillator: Relaxation oscillator → Delayed, unity-gain – FB loop

I_S into C_S ramps v_S (delay) Until v_S reaches v_T (gain)

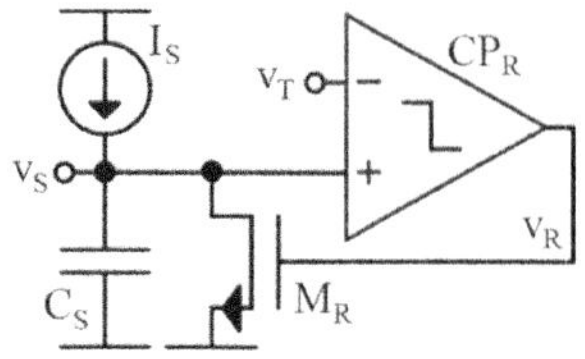

Design: v_S rises past v_T across t_P → $v_{S(HI)} = v_T + \left(\frac{I_S}{C_S}\right)t_P^{+}$

Reset before CP_R trips → $t_R < t_P^{-}$ $\quad t_{CLK} = \left(\frac{C_S}{I_S}\right)\left(v_{S(HI)} - v_{S(LO)}\right) + t_P^{-}$

One-Shot Oscillator: v_I enables oscillations

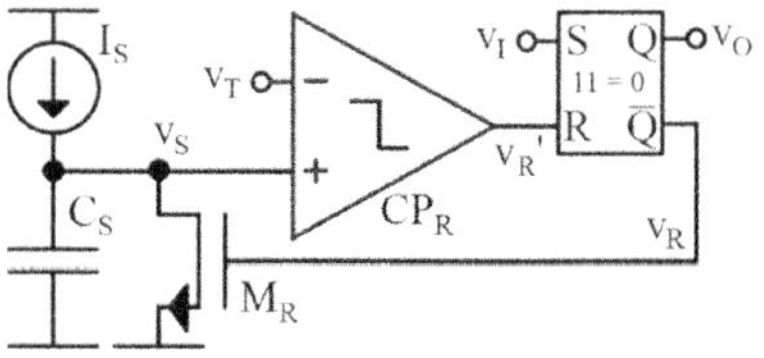

I_S into C_S ramps v_S Until v_S reaches v_T

Design: $v_{S(HI)} = v_T + \left(\frac{I_S}{C_S}\right)\left(t_P^{+} + t_{SR}^{+}\right)$ $\quad t_R < t_P^{-} + t_{SR}^{-} = t_{OFF}$

$t_{PW} = \left(\frac{C_S}{I_S}\right)\left(v_{S(HI)} - v_{S(LO)}\right)$ $\quad t_{OSC} = t_{PW} + t_{OFF}$

Example: Determine v_T, I_S, W_R, & t_{OSC} so $v_{S(HI)} = 300$ mV & $t_{PW} = 750$ ns when

$v_{DD} = 1$ V, $C_S = 5$ pF, $t_P = 100$ ns, $t_{SR} = 1$ ns, $V_{TN0} = 500$ mV, $K_N' = 200\ \mu A/V^2$,

$C_{OX}'' = 6.9\ fF/\mu m^2$, $L_R = 250$ nm, $L_{OL} = 30$ nm, $W_{MIN} = 3\ \mu m$.

Solution: $t_{OFF} = t_P^{-} + t_{SR}^{-} = 100$ ns $\quad t_{OSC} = t_{PW} + t_{OFF} = 850$ ns

$t_R \approx 2.3R_RC_S \approx 2.3R_R(5p) \equiv 25\%t_{OFF} = 25$ ns $\therefore$ $R_R \le 2.2$ kΩ

$$R_R \approx \frac{L_R - 2L_{OL}}{W_RK_N'\left(v_{DD} - V_{TN0}\right)} \le 2.2\ k\Omega \quad \therefore\ W_R \ge 860\ nm \ \rightarrow\ W_R \equiv 3\ \mu m$$

$$v_{S(LO)} \approx -v_{DD}\left(\frac{0.5W_RL_RC_{OX}''}{2C_S}\right) = -520\ \mu V \qquad v_{S(HI)} = v_T + \left(\frac{I_S}{C_S}\right)\left(t_P^{+} + t_{SR}^{+}\right) \equiv 300\ mV$$

$$\therefore\ v_T = 260\ mV$$

$$t_{PW} = \left(\frac{C_S}{I_S}\right)\left(v_{S(HI)} - v_{S(LO)}\right) \equiv 750\ ns \quad \therefore\ I_S = 2\ \mu A$$

8.6. Switch Blocks

Class-A Inverters:

Pull-Down Pull-Up

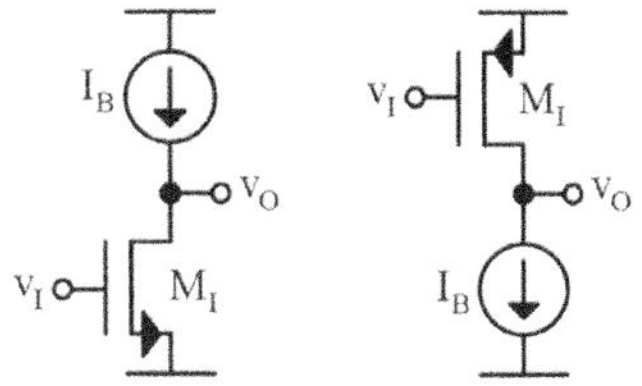

Balanced when $v_{GS} = v_T = V_{GS}\Big|_{I_B}^{V_{DS}>V_{DS(SAT)}} + v_{SS}$

$v_I < v_T$: I_B raises v_O $\quad$ $v_I > v_T$: M_I reduces v_O

Supply-Sensing Comparators: Low Side High Side

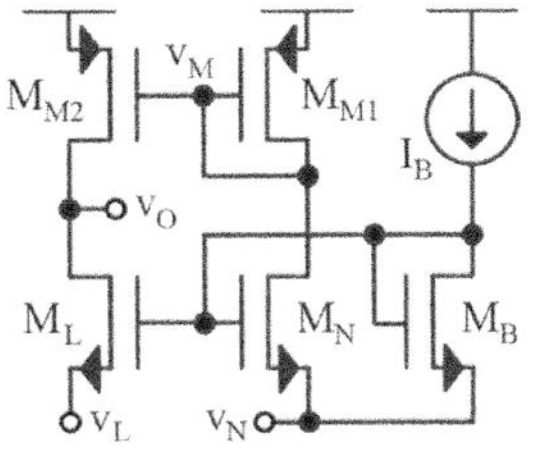

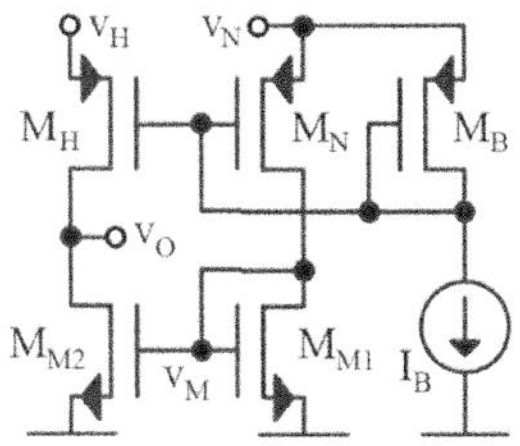

Balanced when $v_L = v_N$

$(V_{GSL} = V_{GSN})$ & $v_O = v_M$

→ $I_L = I_N = I_B$

Create Offset: Unbalance W's or I's

$$V_{OS(S)}\Big|_{V_{DS}>V_{GST}}^{V_{GS}>v_T} = V_{GSL/H} - V_{GSN} \approx \sqrt{\frac{2I_{L/H}}{K'(W/L)_{L/H}}} - \sqrt{\frac{2I_N}{K'(W/L)_N}} \quad \text{If inverted}$$

Zero-Current Detectors:

Low Side:

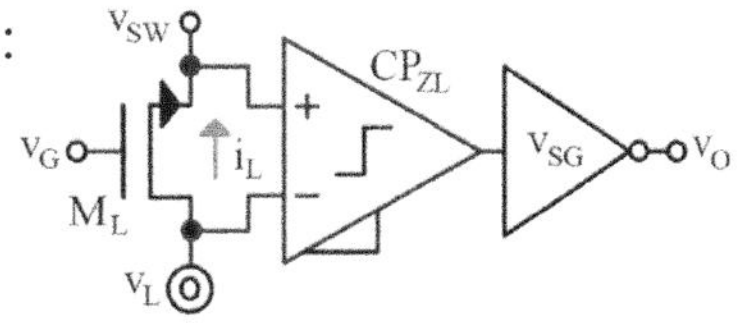

Invoke DCM when i_L reaches zero

Drain switches drain L_X

High Side:

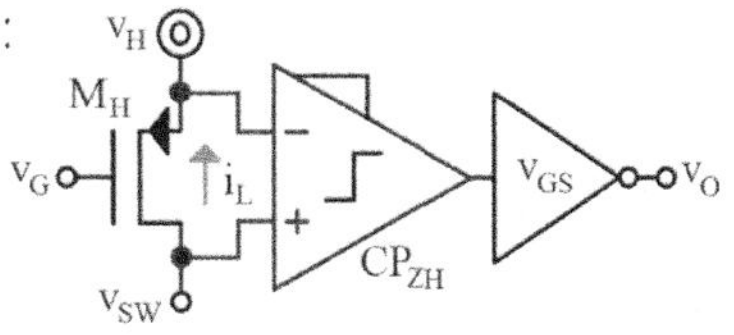

∴ Sense M_{DG} or M_{DO}

v_O trips t_P after i_L crosses zero

∴ Favor $v_{L/H}$ with V_{OS} → Prompt transition before i_L reaches zero

Ring Suppressor:

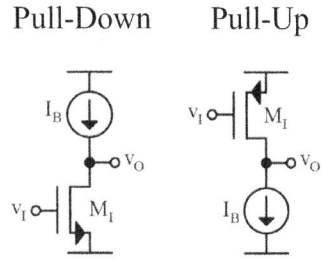

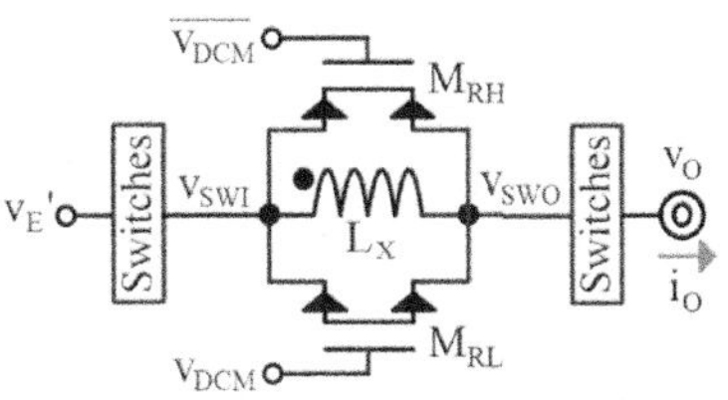

C_{SW}'s & L_X exchange E_{LC} in DCM

Suppress DCM v_{SW} oscillations with R_R

$$t_S = t_{RC} \approx 2.3\tau_{RC} = 2.3R_RC_{SW} < \frac{t_{LC}}{4} = \frac{2\pi\tau_{LC}}{4} = \frac{2\pi/4}{\sqrt{L_XC_{SW}}}$$

v_L approaches zero

∴ Mostly triode R

Switched Diodes: Behave like ideal diodes → $v_D \approx 0$ when $i_D = i_L > 0$

Low Side:

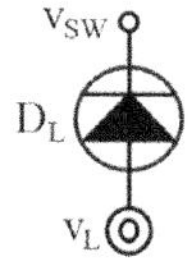

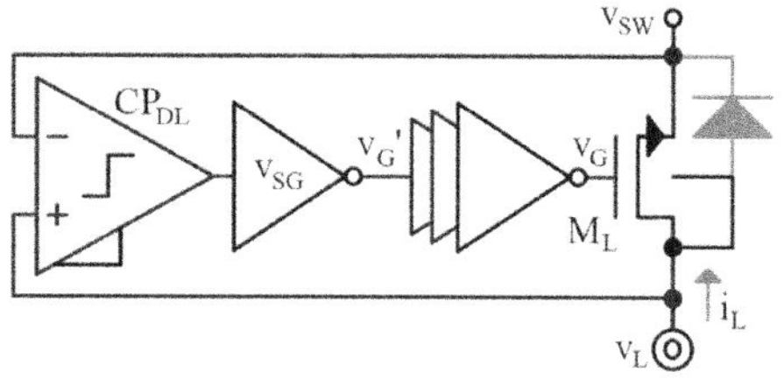

Bucks drain L_X with D_{DG}

M_L switches after t_P

Connect body so body diode drains L_X across t_P

High Side:

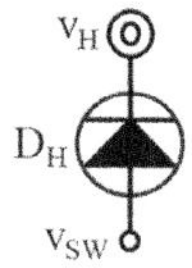

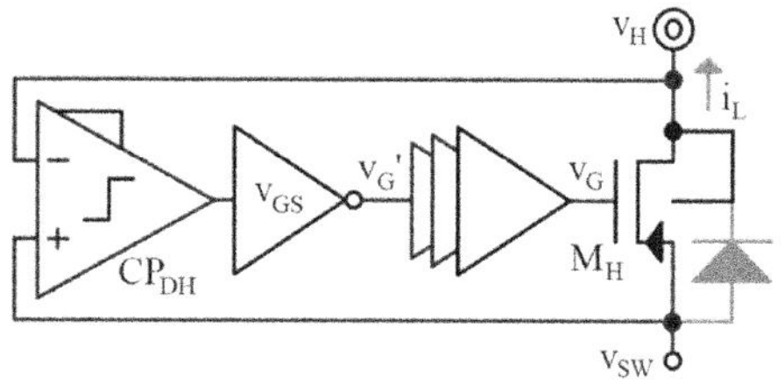

Boosts drain L_X with D_{DO}

Connect body so body diode drains L_X across t_P

Shutdown: Open all switches

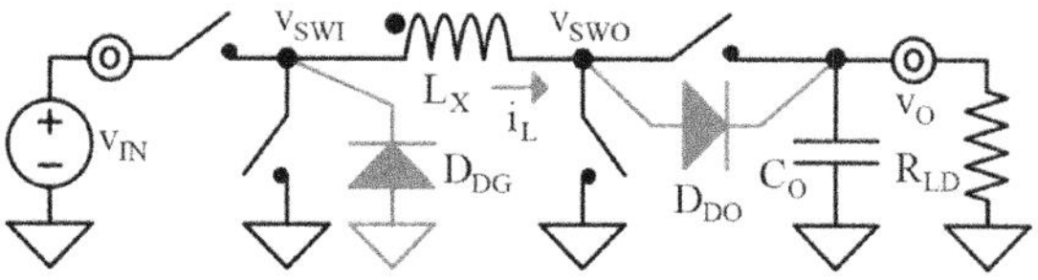

D_{DG} & D_{DO} drain L_X into C_O & R_{LD}

R_{LD} drains C_O

R_L:R_{LD} drain E_{LC} that C_{SW}'s:L_X exchange

Buck–Boost & Buck: v_{SW}'s, v_O, i_L reach zero

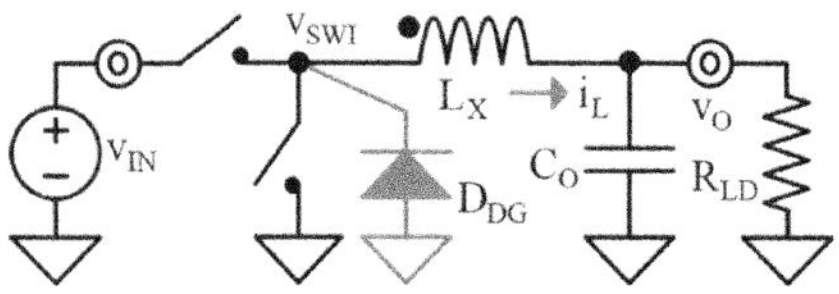

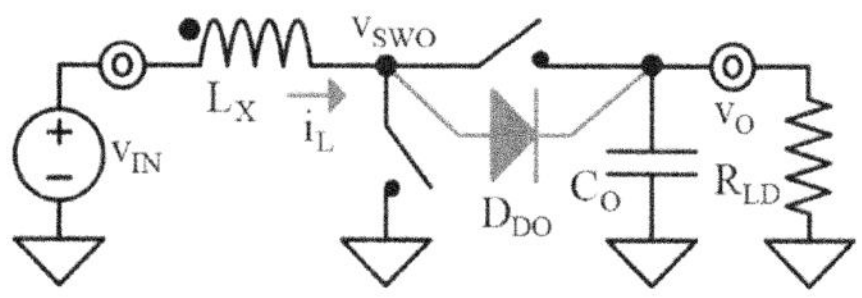

Boost: L_X shorts, $R_{L(DC)}$:R_{LD} load v_{IN} $\therefore\ i_L = \dfrac{v_{IN} - v_{DO}}{R_{L(DC)} + R_{LD}} > 0$ → Disable R_{LD}

Startup: $v_O << \dfrac{v_R}{\beta_{FB}}$ or $i_{L/O} << \dfrac{v_{IR}}{\beta_{IFB}}$ $\therefore$ FB controller raises i_L → Over-shoot can burn components

Starter: Limit i_L or d_E Ramp i_L or d_E Ramp $v_{R/IR}$

Printed in Great Britain
by Amazon

29795552R00071